D1060927

# MATHEMATICAL GEMS III

*By*
ROSS HONSBERGER

THE
DOLCIANI MATHEMATICAL EXPOSITIONS

*Published by*
THE MATHEMATICAL ASSOCIATION OF AMERICA

---

*The Dolciani Mathematical Expositions*

NUMBER NINE

# MATHEMATICAL GEMS III

*By*
ROSS HONSBERGER
*University of Waterloo*

*Published and Distributed by*
THE MATHEMATICAL ASSOCIATION OF AMERICA

*Library of Congress Catalog Card Number 85-061842*

Complete Set ISBN 0-88385-300-0
Vol. 9 ISBN 0-88385-313-2

*Printed in the United States of America*

Current printing (last digit):
10 9 8 7 6 5 4 3 2 1

The DOLCIANI MATHEMATICAL EXPOSITIONS series of the Mathematical Association of America was established through a generous gift to the Association from Mary P. Dolciani, Professor of Mathematics at Hunter College of the City University of New York. In making the gift, Professor Dolciani, herself an exceptionally talented and successful expositor of mathematics, had the purpose of furthering the ideal of excellence in mathematical exposition.

The Association, for its part, was delighted to accept the gracious gesture initiating the revolving fund for this series from one who has served the Association with distinction, both as a member of the Committee on Publications and as a member of the Board of Governors. It was with genuine pleasure that the Board chose to name the series in her honor.

The books in the series are selected for their lucid expository style and stimulating mathematical content. Typically, they contain an ample supply of exercises, many with accompanying solutions. They are intended to be sufficiently elementary for the undergraduate and even the mathematically inclined high-school student to understand and enjoy, but also to be interesting and sometimes challenging to the more advanced mathematician.

---

The following DOLCIANI MATHEMATICAL EXPOSITIONS have been published.

Volume 1: MATHEMATICAL GEMS, by Ross Honsberger

Volume 2: MATHEMATICAL GEMS II, by Ross Honsberger

Volume 3: MATHEMATICAL MORSELS, by Ross Honsberger

Volume 4: MATHEMATICAL PLUMS, edited by Ross Honsberger

Volume 5: GREAT MOMENTS IN MATHEMATICS (BEFORE 1650), by Howard Eves

Volume 6: MAXIMA AND MINIMA WITHOUT CALCULUS, by Ivan Niven

Volume 7: GREAT MOMENTS IN MATHEMATICS (AFTER 1650), by Howard Eves

Volume 8: MAP COLORING, POLYHEDRA, AND THE FOUR-COLOR PROBLEM, by David Barnette

# PREFACE

The technical background required for the enjoyment of the essays in this collection seldom goes beyond the level of the college freshman. It is remarkable how much exciting mathematics exists at this elementary level. While we can't help learning something from each new problem, these topics are presented solely for your *enjoyment*, and although a certain amount of drive and concentration is a price that cannot be avoided, I hope you will find that these little gems are well worth the effort necessary for their appreciation.

A glossary of technical terms and ideas is given at the end of the essays; the words that are explained there are marked in the text with an asterisk (*).

I would like to take this opportunity to thank the members of the Subcommittee on the Dolciani Expositions—Joe Malkevitch, Kenneth Rebman, Alan Tucker, and Donald Small—for a careful review of the manuscript; their constructive criticism led to many improvements.

Ross Honsberger

# CONTENTS


CHAPTER PAGE

1. Gleanings from Combinatorics .... 1
2. Gleanings from Geometry .... 18
3. Two Problems in Combinatorial Geometry .... 36
4. Sheep Fleecing with Walter Funkenbusch .... 48
5. Two Problems in Graph Theory .... 56
6. Two Applications of Generating Functions .... 64
7. Some Problems from the Olympiads .... 76
8. A Second Look at the Fibonacci and Lucas Numbers .... 102
9. Some Problems in Combinatorics .... 139
10. Four Clever Schemes in Cryptography .... 151
11. Gleanings from Number Theory .... 174
12. Schur's Theorem: An Application of Ramsay's Theorem .... 183
13. Two Applications of Helly's Theorem .... 186
14. An Introduction to Ramanujan's Highly Composite Numbers .... 193
15. On Sets of Points in the Plane .... 201
16. Two Surprises from Algebra .... 208
17. A Problem of Paul Erdös .... 215
18. Cai Mao-Cheng's Solution to Katona's Problem on Families of Separating Subsets .... 224

Solutions to Selected Exercises .... 241
Glossary .... 247
Index .... 249

CHAPTER 1

# GLEANINGS FROM COMBINATORICS

The best source of elementary problems in mathematics is undoubtedly the many annual contests that have sprung up all around the world. Besides the Putnam Competition [1] and the International Olympiad, there must be at least a dozen national olympiads. The questions that are submitted for these contests are screened so carefully that the final slate of problems usually contains at least a couple of real gems. They are particularly appealing because their solutions generally turn more on ingenuity than special knowledge or manipulation. Many of these contests are published in the outstanding problems journal *Crux Mathematicorum* [2]. We shall encounter several problems from the olympiads throughout this book. To begin, let us consider a problem from a recent Putnam paper.

## 1. A Putnam Paper Problem (Problem A4, 1979)

> Let any $2n$ points be chosen in the plane so that no 3 are collinear, and let any $n$ of them be colored red and the other $n$ blue. Prove that it is always possible to pair up the $n$ red points with the $n$ blue ones in 1–1 fashion $(r, b)$ so that no 2 of the $n$ segments $rb$, which connect the members of a matched pair, intersect.

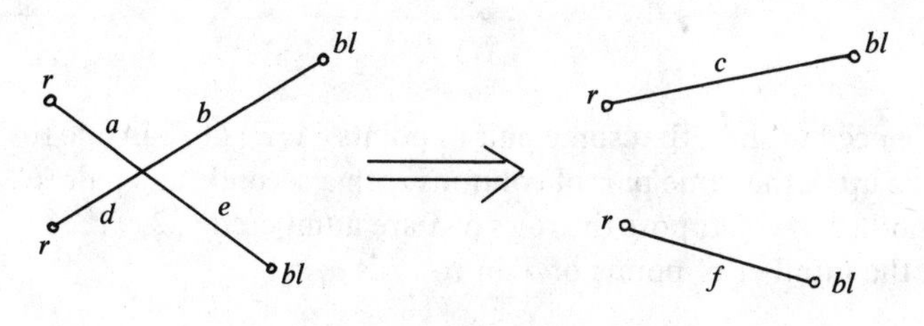

Since $n$ is finite, there exists altogether only a finite number of ways of pairing up the sets of red and blue points. For each way of doing this, the lengths of the $n$ segments add up to some total amount, which is generally different for different sets of matchings. However, if an arrangement contains an intersection, the following argument, based on the triangle inequality, shows that there exists an almost identical arrangement which has a smaller total length: undo the intersection as shown, keeping the other $n - 2$ segments unchanged ($a + b > c, d + e > f$). Consequently, any arrangement which has minimum total length must necessarily be free of intersections.

## 2. Two Problems from the 1974 USSR National Olympiad (#4 and #9)

#4. Consider a square grid $S$ of 169 points which are uniformly arrayed in 13 rows and 13 columns (like the lattice points $(m, n), m, n = 1, 2, \ldots, 13$). Prove that no matter what subset $T$, consisting of 53 of these points, might be selected, some 4 points of $T$ will be the vertices of a rectangle $R$ whose sides are parallel to the sides of $S$.

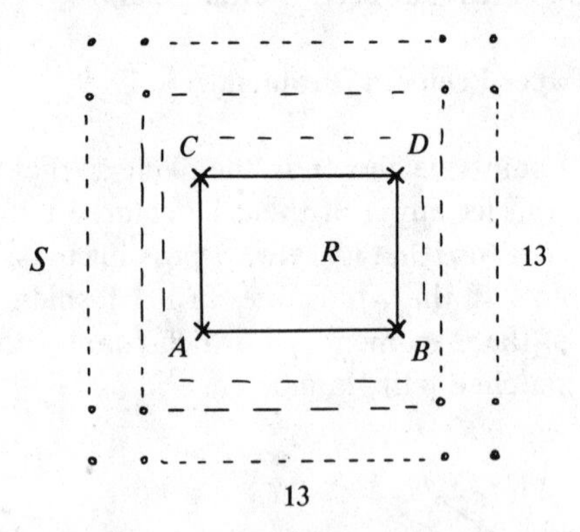


We need to show that some pair of points $(A, B)$ of $T$, in one row of $S$, line up in the same pair of columns with a second such pair $(C, D)$ in another row. Suppose the rows of $S$ are numbered $1, 2, \ldots, 13$ and that the number of points of $T$ in row $i$ is $a_i$.

Now $a_i$ points in the same row determine $\binom{a_i}{2}$ pairs of candidates $(A, B)$, each occurring in one of the $\binom{13}{2}$ possible pairs of columns of $S$. Of course, there is a very good chance that none of the $\binom{a_i}{2}$ pairs of columns determined by the points of $T$ in a particular row $i$ will occur again among the $\binom{a_j}{2}$ such pairs for any other row $j$. But, if the *total* number of pairs of columns determined by the rows of points of $T$, namely $\binom{a_1}{2} + \binom{a_2}{2} + \cdots + \binom{a_{13}}{2}$, were to exceed $\binom{13}{2}$, the number of possible pairs of columns of $S$, the pigeonhole principle* would imply that some pair of columns would have to be repeated, and thus produce a desired rectangle $R$. Therefore, let us try to show that

$$\sum_{i=1}^{13} \binom{a_i}{2} > \binom{13}{2},$$

which simplifies easily as follows (since $T$ contains 53 points, we have $\Sigma_{i=1}^{13} a_i = 53$):

$$\sum_{i=1}^{13} \frac{a_i(a_i - 1)}{2} > \frac{13 \cdot 12}{2},$$

$$\sum_{i=1}^{13} (a_i^2 - a_i) > 13 \cdot 12,$$

$$\sum_{i=1}^{13} a_i^2 > 156 + \sum_{i=1}^{13} a_i = 156 + 53 = 209.$$

By setting each $b_i = 1$ in the famous Cauchy inequality*

$$(a_1^2 + a_2^2 + \cdots + a_n^2)(b_1^2 + b_2^2 + \cdots + b_n^2) \geq (a_1 b_1 + a_2 b_2 + \cdots + a_n b_n)^2$$

(which, I expect, would be well known to all the olympiad mathletes, who are thoroughly coached these days), we obtain

$$\left(\sum_{i=1}^{13} a_i^2\right) \cdot 13 \geq \left(\sum_{i=1}^{13} a_i\right)^2,$$

* An asterisk indicates a word or idea that is explained in the glossary.

which gives

$$\sum_{i=1}^{13} a_i^2 \geq \frac{53^2}{13} > 53 \cdot \frac{52}{13} = 53 \cdot 4 = 212 > 209,$$

as desired.

This companion problem approaches the same subject from the opposite point of view, a nice touch by the composers of this olympiad.

> #9. Given a square grid $S$ containing 49 points in 7 rows and 7 columns, a subset $T$ consisting of $k$ points is selected. The problem is to find the maximum value of $k$ such that no 4 points of $T$ determine a rectangle $R$ having sides parallel to the sides of $S$.

Using the notation established in the previous problem, we see immediately that unless

$$\sum_{i=1}^{7} \binom{a_i}{2} \leq \binom{7}{2},$$

an undesired rectangle $R$ will surely result. Since $\Sigma_{i=1}^{7} a_i = k$, this reduces to

$$\sum_{i=1}^{7} a_i^2 \leq 42 + k \tag{1}$$

Turning again to the Cauchy inequality, putting each $b_i = 1$ yields

$$(a_1 + a_2 + \cdots + a_7)^2 \leq \left(\sum_{i=1}^{7} a_i^2\right) \cdot 7,$$

and

$$\frac{k^2}{7} \leq \sum_{i=1}^{7} a_i^2 \tag{2}$$

Combining (1) and (2), we obtain

$$\frac{k^2}{7} \leq \sum_{i=1}^{7} a_i^2 \leq 42 + k,$$

which demands that

$$\frac{k^2}{7} \le 42 + k,$$

$$k^2 - 7k - 294 \le 0,$$

$$(k + 14)(k - 21) \le 0,$$

placing $k$ in the range $-14 \le k \le 21$.

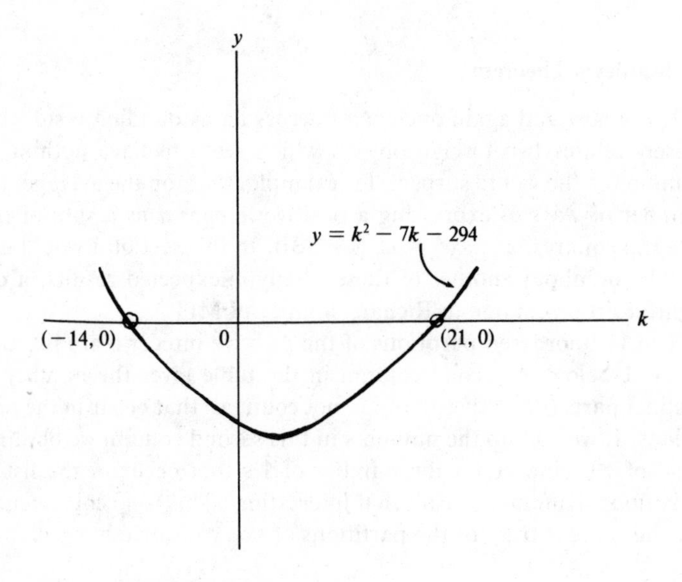


Is it possible for $k$ to be as large as 21? If $k = 21$, all of the inequalities, including the Cauchy inequality, become equalities. Now there is equality in Cauchy's relation only if the $a_i$ and $b_i$ are, respectively, proportional. Since all $b_i = 1$, equality here is out of the question unless the $a_i$'s are all equal. Thus, if $k$ can actually be as great as 21, $T$ will have to contain exactly 3 points from each row. In checking the feasibility of such an arrangement by direct trial, one soon succeeds as shown. Thus the maximum $k$ is indeed 21.

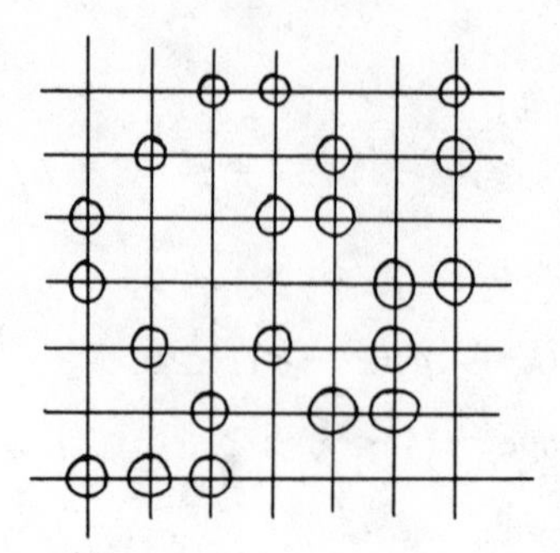

### 3. Stanley's Theorem

Every now and again one comes across an astounding result that closely relates two foreign objects which seem to have nothing in common. Who would suspect, for example, that, on the average, the number of ways of expressing a positive integer $n$ as a sum of two integral squares, $x^2 + y^2 = n$, is $\pi$ ([**3**]). In this section I would like to tell you about another of these totally unexpected results, a delightful little gem due to Richard Stanley of MIT.

The 11 unordered partitions of the positive integer 6 are listed in Table 1 below. A second column in the table gives the number of distinct parts (i.e., repetitions are not counted) that occur in the partitions. If we add up the numbers in this second column we obtain a total of 19. Now count the number of 1's that occur in the list of partitions. Hmmm . . . isn't that interesting? This is no coincidence, for the same is true for the partitions of every positive integer.

STANLEY'S THEOREM. *The total number of* 1*'s that occur among all unordered partitions of a positive integer is equal to the sum of the numbers of distinct parts of those partitions.*

Let us denote the number of unordered partitions of $n$ by $p(n)$, and define $p(0)$ to be 1. Since the order of the integers in a partition doesn't count, rearranging them to suit ourselves doesn't cause any trouble. Consequently, let us write them in nondecreasing order and enter them in a table (in a normal way—one partition per row, starting each at the left). Each partition will occupy as many columns as it

| Partition | Number of Distinct Parts |
|---|---|
| 6 | 1 |
| 5 + 1 | 2 |
| 4 + 2 | 2 |
| 4 + 1 + 1 | 2 |
| 3 + 3 | 1 |
| 3 + 2 + 1 | 3 |
| 3 + 1 + 1 + 1 | 2 |
| 2 + 2 + 2 | 1 |
| 2 + 2 + 1 + 1 | 2 |
| 2 + 1 + 1 + 1 + 1 | 2 |
| 1 + 1 + 1 + 1 + 1 + 1 | 1 |
| TOTAL | 19 |

Table 1

has parts, the rest of its row remaining empty. In order to accommodate the partition which consists entirely of 1's, the table will have to extend to $n$ columns. It is our desire to count the total number of 1's that are present in the table. We shall do this by determining the number of 1's that occur in column $k$ and summing as $k$ goes from 1 to $n$.

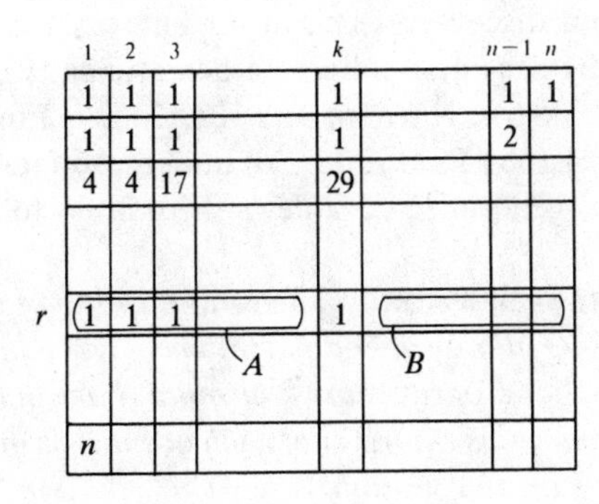


Table 2

The question is, then, when does a partition place a 1 in column $k$? Let us denote by $A$ the section of a partition that precedes column $k$ and by $B$ the section which follows column $k$ (see row $r$ in Table 2). Because the partitions are written in nondecreasing order, a 1 in column $k$ means that every entry in $A$ is also a 1 and that the parts in the first $k$ columns of the partition (being all 1's) add up to $k$. This implies that the entries in $B$ must add up to $n - k$. Conversely, if any of the $p(n - k)$ partitions of the integer $n - k$ were to be combined with $k$ 1's, a partition of $n$ would result that would place a 1 in column $k$ of our table. Therefore, the number of 1's in column $k$ is $p(n - k)$, and the total number of 1's in the table is $\Sigma_{k=1}^{n} p(n - k)$. (When $k = n$, this involves $p(0)$, which we have defined to be 1 in order to count the single 1 that occurs in column $n$ of the table.)

In order to determine the sum of the numbers of distinct parts, let us list beside each partition its set of distinct parts, as in Table 3. The number in question is the total *number* of integers that occur in the second part of the table. We can determine how many numbers are here by finding out how many 1's there are, how many 2's, and so on. Let us consider how many times the integer $k$ occurs in this section of the table (see row $s$). If $k$ is among the parts of the partition, the rest of the partition (denoted by $X$ and $Y$ in the table, and which may or may not contain more $k$'s) must add up to $n - k$. Conversely, if the integer $k$ is combined with any of the $p(n - k)$ partitions of $n - k$, a partition of $n$ is obtained which has $k$ among its parts. Hence the integer $k$ occurs $p(n - k)$ times in the second section of Table 3, and the sum of the numbers of distinct parts is $\Sigma_{k=1}^{n} p(n - k)$, proving Stanley's fabulous discovery. (This argument was derived from independent proofs by two outstanding mathematicians from the Netherlands—E. W. Dijkstra, Nuenen, and K. A. Post, Eindhoven.)

In the spring of 1984 Paul Elder, an undergraduate at the University of Waterloo, generalized Stanley's theorem as follows.

ELDER'S GENERALIZATION. *The total number of occurrences of an integer k among all unordered partitions of n is equal to the number of occasions that a part occurs k or more times in a partition. (A partition which contains r parts that each occur k or more times contributes r to the sum in question.)*

| | Partition | Distinct Parts |
|---|---|---|
| | $1+1+\cdots+1$ | 1 |
| | $2+3+5+\cdots$ | 2, 3, 5, $\cdots$ |
| | | |
| $s$ | $\boxed{X} + k + \boxed{Y}$ | $\cdots$, k, $\cdots$ |
| | | |
| | $n$ | $n$ |

Table 3

For example, in the partitions of 6, above, the integer 2 occurs 8 times, and there are 8 occasions of a partition containing a repeated part ($2 + 2 + 1 + 1$ containing two such occasions).

*Proof.* The number of occasions in which an integer $i$ occurs $k$ or more times in a partition is $p(n - ik)$ (which is just the number of ways of partitioning the remaining $n - ik$ units of the number, allowing for additional parts equal to $i$). The total number of occasions $T$ in which a part occurs $k$ or more times, then, is obtained as $i$ runs from 1 to $n$:

$$T = p(n - k) + p(n - 2k) + p(n - 3k) + \cdots + p(n - nk).$$

(We define $p(m)$ to be 0 for $m$ negative.)

On the other hand, $p(n - ik)$ is also the number of partitions in which $k$ occurs $i$ or more times. Consequently, a partition which contains exactly $r$ $k$'s will be counted by each of the $r$ terms

$$p(n - k), p(n - 2k), \ldots, p(n - rk),$$

but not by any of the later terms in $T$. That is to say, such a partition is counted $r$ times in $T$, and it follows that $T$ also gives the number of times $k$ occurs among all the partitions.

## 4. A Surprising Partition

A problem is usually highly rated because it either raises an interesting challenge, announces an intriguing result, or has an ingenious solution. We consider next a problem which may be pretty short on a couple of these counts but which has the unusual feature of a marvellous *answer*. This matter was brought to my attention by David Shapiro of Ohio State University.

> Determine how to partition a positive integer $n > 1$ into one or more positive integers,
>
> $$n = m_1 + m_2 + \cdots + m_k,$$
>
> so that the product of the parts $P = m_1 m_2 \cdots m_k$ is a maximum.

One's first impulse is to try to make the parts as nearly equal as possible. For $n = 23$, for example, we observe successive improvements in the product $P$ through the partitions $23 = 11 + 12 = 7 + 8 + 8 = 5 + 6 + 6 + 6 = \cdots$ (with products 23, 132, 448, 1080, ...), but disappointment looms on the horizon in the final such partition, $1 + 1 + 1 + \cdots + 1$, with $P = 1$. Notwithstanding this final setback, it turns out, in retrospect, that the notion of equal parts is ultimately a good one. However, let us not digress any further along this line.

As in many cases, the desired general rule can be stated in more than one way. It strikes me as most intriguing that the answer to our problem can be stated in the following form: TAKE AS MANY 3's AS POSSIBLE, using only 2's and 3's. For example, 20 gives six 3's and a 2, 21 gives seven 3's, while 22 gives six 3's and two 2's. Why 3's? Is it because 3 is the integer nearest $e$? In any event, once the secret is out, the difficulties soon evaporate. The easy proof is left to the reader. (Hint: if a part $m \geq 5$, peel off 3 to yield 3 and $m - 3$; then the contribution to the product is $3(m - 3) = m + (2m - 9)$, which exceeds $m$ because $m \geq 5$.)

## 5. An International Olympiad Problem

The second problem on the first section of the 1979 International Olympiad proposes a rather surprising result.

> The convex* pentagons $A = A_1A_2A_3A_4A_5$ and $B = B_1B_2B_3B_4B_5$ lie in different planes and each vertex of $A$ is joined by a segment to each vertex of $B$. Suppose that each of the edges of $A$ and $B$ and the 25 segments $A_iB_j$ (that is, every possible edge except the diagonals of $A$ and $B$) is colored either red or blue so that no triangle formed by 3 of these segments has all its edges the same color (i.e., is monochromatic). Prove, then, that the 10 edges of $A$ and $B$ must all be the same color.

Let us show first that $A$ and $B$ must each be monochromatic and then that they must both be the same color.

(i) We proceed indirectly. Suppose that $A$ is not monochromatic. In this case, as we go around $A$ from edge to edge, we must come upon two consecutive edges that have different colors. Without loss of generality, we may suppose that $A_1A_2$ is red and $A_2A_3$ is blue. Now there are 5 colored segments from $A_2$ to the vertices of $B$. Since

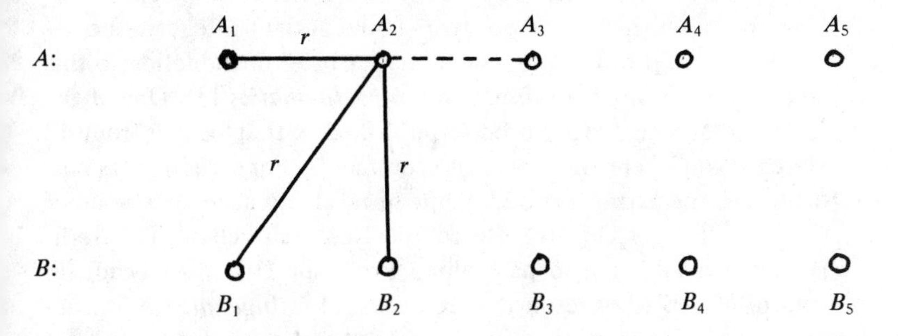


there are only 2 colors, the pigeonhole principle implies that 3 of these segments must be the same color, say red. Because $B$ has only 5 vertices, 2 of these 3 red segments must go to a pair of *consecutive* vertices of $B$, say $B_1$ and $B_2$ as shown (although our figure does not indicate it, $B_1$ and $B_5$ are consecutive vertices of $B$). Now then, we have the "squeeze" on triangle $A_1B_1B_2$. Each of its edges, uncolored in the figure, belongs to a triangle having 2 red edges (e.g., $A_1B_2$ is in triangle $A_1A_2B_2$). Therefore, if any edge of triangle $A_1B_1B_2$ is colored red, a monochromatic red triangle is completed; otherwise

$A_1B_1B_2$ itself will be a monochromatic blue one. In every case, then, a forbidden monochromatic triangle is forced to arise, giving a contradiction. Thus $A$, and similarly $B$, must be monochromatic.

(ii) Suppose, however, that $A$ is all red and $B$ is all blue. Now, as we just observed, if a vertex $A_i$ were to send out as many as 3 blue edges to the vertices of $B$, 2 of them would go to a pair of consecutive vertices. Since $B$ is assumed to be all blue, a monochromatic triangle results. Consequently, each vertex $A_i$ must send not more than 2 blue edges to $B$, for a total of 10 or fewer blue edges among the $A_iB_j$.

Similarly, no vertex of $B$ could send more than 2 red edges to an all-red $A$, implying a total of not more than 10 red edges among the $A_iB_j$. Altogether, then, there could not be more than 20 edges $A_iB_j$, in contradiction to the actual number of 25, and the argument is complete.

## 6. Complementary Sequences

If the union of two disjoint sequences of positive (nonnegative) integers is the entire set of positive (nonnegative) integers, the sequences are said to be complementary. A brief introduction to this topic is given in my book *Ingenuity in Mathematics* [**3**]. One of the nice things about having a book published is that people from all over send you interesting notes and comments. For several years now I have had the extreme good fortune of receiving a steady stream of gems from Edsger Dijkstra, Burroughs Research Fellow, The Netherlands. I would like to pass along Professor Dijkstra's beautiful proof of a surprising result that is discussed in *Ingenuity in Mathematics.* In order to do this, it will be necessary to undertake a brief review of the basic notions.

Suppose $f(n)$, $n = 0, 1, \ldots$, is a nondecreasing sequence of nonnegative integers which is unbounded above (for example, the Fibonacci sequence, defined by $f(0) = 0, f(1) = 1$ and, for $n > 1$, $f(n) = f(n-1) + f(n-2)$). From $f(n)$ we can derive a second function $g(n)$ as follows: $g(n) =$ the number of values $m$ for which $f(m) \leq n$; thus $g(0)$ is the number of times $f(m) \leq 0$, $g(1)$ is the number of times $f(m) \leq 1$, and so on (see the table).

| $n$ | 0 | 1 | 2 | 3 | 4 | 5 | 6 | 7 | 8 | 9 | 10 | $\cdots$ |
|---|---|---|---|---|---|---|---|---|---|---|---|---|
| $f(n)$ | 0 | 1 | 1 | 2 | 3 | 5 | 8 | 13 | 21 | 34 | 55 | $\cdots$ |
| $g(n)$ | 1 | 3 | 4 | 5 | 5 | 6 | 6 | 6 | 7 | 7 | 7 | $\cdots$ |
| $F(n)$ | 0 | 2 | 3 | 5 | 7 | 10 | 14 | 20 | 29 | 43 | 65 | $\cdots$ |
| $G(n)$ | 1 | 4 | 6 | 8 | 9 | 11 | 12 | 13 | 15 | 16 | 17 | $\cdots$ |

If $f(n)$ were to remain constant for $n \geq N$, then $g(n)$ would be infinite for $n \geq N$. In order to avoid this, we have insisted that $f(n)$ be unbounded. To our mild surprise, we always find that $f(n)$ and $g(n)$ are *inverse* sequences, that is, $g(n)$ is also a nondecreasing unbounded sequence of nonnegative integers, and if we had begun with $g(n)$ in the first place, the second function would have turned out to be $f(n)$. We shall establish this shortly.

From $f(n)$ and $g(n)$ we construct our main interests, the functions $F(n)$ and $G(n)$, as follows:

$$F(n) = f(n) + n, \qquad G(n) = g(n) + n$$

(see the table). Then, to our astonishment, no matter what sequence $f(n)$ one begins with, the resulting sequences $F(n)$ and $G(n)$ are always a pair of complementary sequences! Professor Dijkstra proves this with the aid of graphs as follows.

Suppose we draw a bar graph of $f(n)$, as illustrated in Figure 1, in which the value $f(n) = k$ is represented by a unit segment joining $(n, k)$ to $(n + 1, k)$. The segments parallel to the $y$-axis are inserted merely to guide the eye and are not really essential parts of the graph. However, they serve us in an unexpected way as follows. If one moves along the line $y = n$ in a positive direction from the $y$-axis, the distance to the vertical segment of the graph which extends above this line is the value of $g(n)$ (Figure 2): a horizontal span of length $k$, for example, indicates that, for the $k$ values $m = 0, 1, \ldots, k - 1$, we have $f(m) \leq n$, while $f(k) > n$. That is to say, the sequence $g(n)$ is automatically recorded relative to the $y$-axis when $f(n)$ is plotted against the $x$-axis. In passing, we note that this gives an immediate

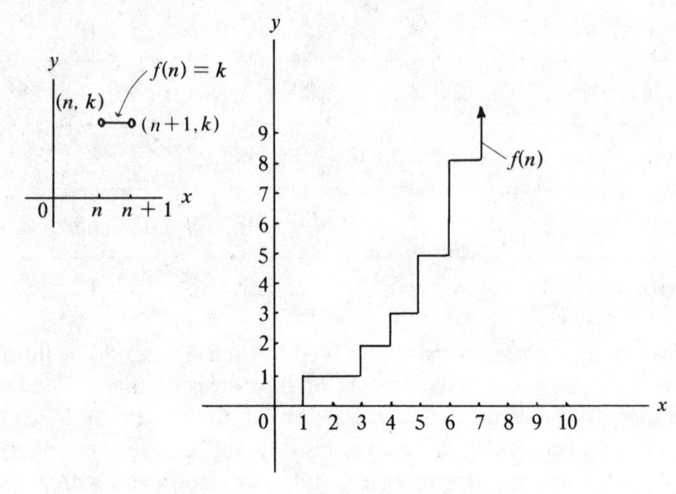


Figure 1

proof that $f(n)$ and $g(n)$ are inverse sequences: since the graphs complement each other in the first quadrant, plotting either against the $x$-axis produces the other along the $y$-axis.

Now let us extend our bar graph as follows (Figure 3): for each $n$, let the vertical bar which represents the value $f(n)$ be extended down below the $x$-axis a distance of $n$ units to form a bar of total length $f(n) + n = F(n)$. Similarly, let each horizontal bar which represents $g(m)$ be extended into the second quadrant a distance $m$ to give a bar of total length $g(m) + m = G(m)$. Thus we obtain a bar graph of the values of $F(n)$ and $G(n)$ in the form of an inverted staircase that extends indefinitely along the line $y = -x$. In order to prove that $F(n)$ and $G(n)$ are complementary sequences, we must show that our staircase contains exactly one bar of each of the lengths 0, 1, 2, . . . . We can accomplish this very neatly by tracing the graph of $f(n)$ as it snakes through the first quadrant, separating the bars which give the values of $F$ and $G$.

Let us trace $f(n)$ using steps of unit length. The first step must carry us from the origin along the $x$-axis to (1, 0) (if $f(0) = 0$) or up the $y$-axis to (0, 1) (if $f(0) > 0$), and, in either case, it cuts from the staircase a bar of size 0.

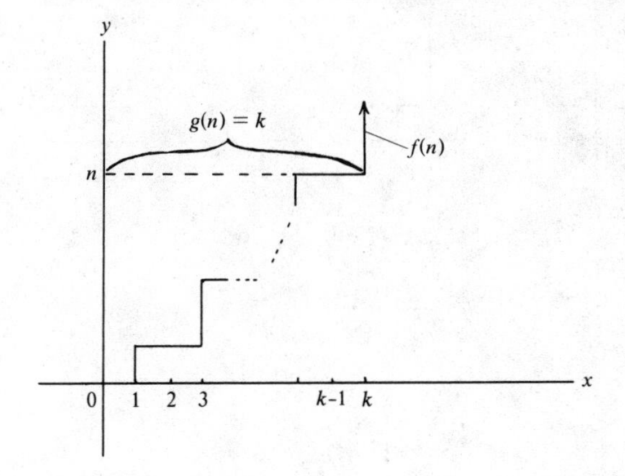


Figure 2

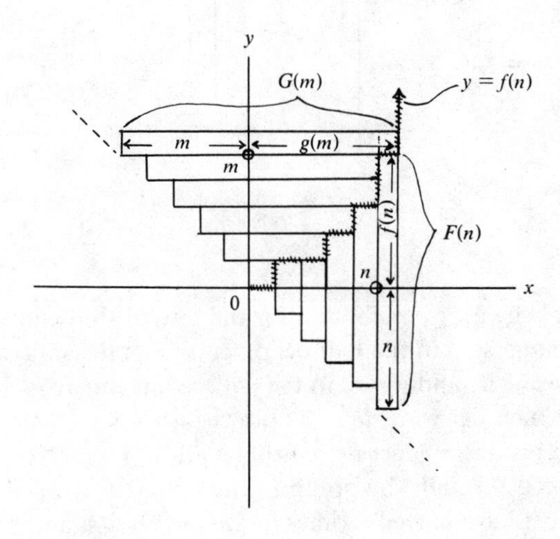


Figure 3

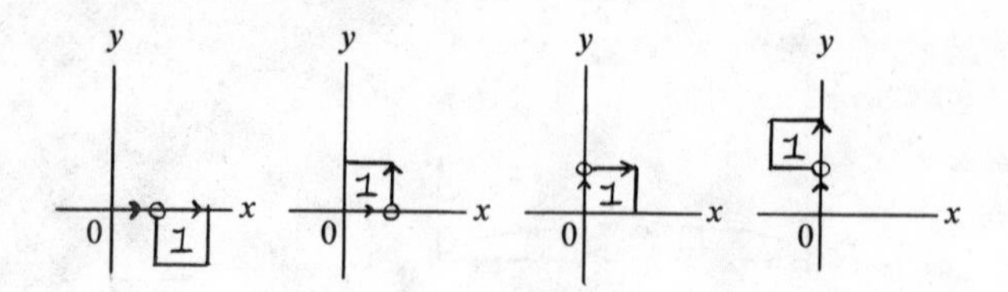


Figure 4. Step two.

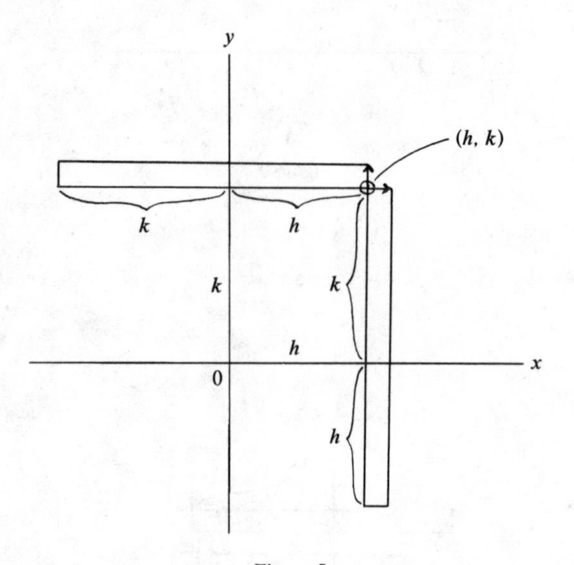


Figure 5

There are only four possibilities for the second step and it is clear in general that a step in the $x$-direction cuts from the staircase a bar which lies *below* it, and a step in the $y$-direction cuts from the staircase a bar which lies to its *left*. By direct inspection, then, we have that the first two steps generate bars of lengths 0 and 1. Now suppose we have traced through the graph to the point $(h, k)$. Taking unit steps, we must have moved $h$ times in the $x$-direction and $k$ times in the $y$-direction, for a total of $h + k$ steps. Suppose that these steps have cut from the staircase bars of lengths $0, 1, 2, \ldots, h + k - 1$. From Figure 5 it is obvious that the vertical and horizontal bars at $(h, k)$ are by definition extended below the $x$-axis a distance $h$ and to the left of the $y$-axis a distance $k$. The next step must necessarily determine a bar of length $h + k$; the desired conclusion thus follows by induction.

### References

1. A. M. Gleason, R. E. Greenwood, and L. M. Kelly, The William Lowell Putnam Mathematical Competition: Problems and Solutions: 1938–1964, Mathematical Association of America, 1980.
2. Crux Mathematicorum, Algonquin College, Ottawa, Ontario, Canada.
3. R. Honsberger, Ingenuity in Mathematics, New Mathematical Library Series, vol. 23, Mathematical Association of America, 1970.

CHAPTER **2**

# GLEANINGS FROM GEOMETRY

## 1. Morsel #23

In my book *Mathematical Morsels* [**1**], two very nice solutions are given to Morsel #23. However, the following exciting variation of the first one takes the prize. It was most kindly sent to me by Professor William Moser of McGill University.

> Morsel #23. A set $S$ of $2n + 3$ points is given in the plane, no 3 on a line and no 4 on a circle. Prove that it is always possible to find a circle $C$ which goes through exactly 3 points of $S$ and splits the other $2n$ in half, that is, $n$ on the inside and $n$ on the outside.

Let a straight line $L$, initially far to one side of $S$, sweep across the plane until it passes through some point of $S$, say $A$. Then let $L$ be rotated about $A$ until it goes through a second point $B$ of $S$ (in more sophisticated language, let $L$ be a side of the convex hull* of $S$). Since no 3 points of $S$ lie on a straight line, $A$ and $B$ will be the only points of $S$ on $L$, and the other $2n + 1$ points of $S$, which we call $P_1$, $P_2$, ..., $P_{2n+1}$, will all lie on the same side of $L$ (see the figure).

Now if the segment $AB$ were to subtend the same angle at two of these points $P_i$ and $P_j$, the 4 points $A$, $B$, $P_i$, $P_j$ would lie on a circle. Since this is forbidden, it must be that $AB$ subtends a different angle at each of the points $P_i$. Let us suppose that the names $P_i$ have been assigned so that the angle at $P_1$ is the smallest, the angle at $P_2$ the next in size, and so on to the largest angle at $P_{2n+1}$. Because there is an odd number of these angles, there must be one in the middle, namely, the angle at $P_{n+1}$. Now it is clear that the circle $C$ through $A$, $B$, and $P_{n+1}$ solves our problem: since $AB$ subtends the same angle at all points on the arc $AP_{n+1}B$ of this circle, the vertices $P_1$, $P_2$,

$\ldots, P_n$ of the $n$ smaller angles must all lie outside $C$, while those of the $n$ larger angles, $P_{n+2}, P_{n+3}, \ldots, P_{2n+1}$, must all lie inside $C$.

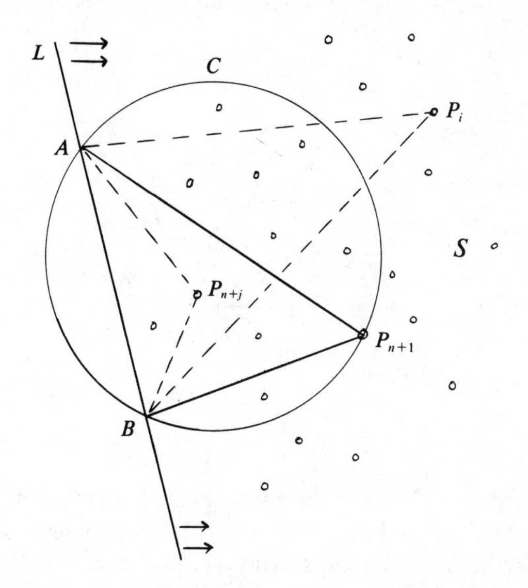


This construction also provides an indication of the number of "solution" circles that exist. It yields a circle for each side of the convex hull of $S$. However, it is conceivable that one could get the same circle more than once—if either $AP_{n+1}$ or $BP_{n+1}$ is also a side of the convex hull. In any case, no circle could be obtained by this construction more than twice, for not all 3 sides of triangle $ABP_{n+1}$ could belong to the convex hull. Thus, if the convex hull is a $k$-gon, there must be at least $k/2$ solution circles.

## 2. Mrs. Dijkstra

In the course of my correspondence with Professor Edsger Dijkstra (Nuenen, The Netherlands), I sent along an essay on geometry in which the following problem was mentioned, without solution, as a simple exercise:

> Around equilateral triangle $ABC$ circumscribe a rectangle in any direction you like (say $PBQR$). In general, each

side of $ABC$ cuts off a right triangle from the rectangle. Prove that the areas of the two smaller triangles always add up to the area of the largest one ($X = Y + Z$).

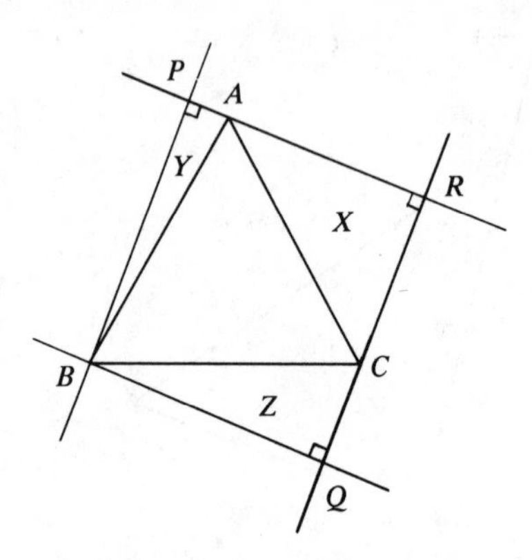


Professor Dijkstra knew that his mother was fond of geometry and he thought that she might have some fun with this problem. So he sent it to her (now Mrs. B. C. Dijkstra-Kluyver, Zutphen, The Netherlands) and, in return mail, he got back the following solution.

Since $AB = AC$ ($ABC$ is equilateral), let triangle $PAB$ be rotated about $A$, across $\Delta ABC$, until $AB$ coincides with $AC$. Similarly, suppose triangle $QCB$ is rotated about $C$, across $\Delta ABC$, until $BC$ coincides with $AC$. Let the new positions of $P$ and $Q$ be $P'$ and $Q'$ (see the figure). Since $X$, $Y$, and $Z$ now have a common base $AC$, the areas of $Y$ and $Z$ will add up to the area of $X$ if and only if the altitudes to $AC$ from $P'$ and $Q'$ add up to the altitude from $R$.

Because $AC$ subtends right angles at $P'$, $Q'$, and $R$, the circle $K$ on $AC$ as diameter will pass through each of these points. Now, in the original figure the angle $RAP$ was $180°$. However, it was reduced by $60°$ by the rotation about $A$ ($\langle BAC = 60°$ in equilateral triangle $ABC$). Consequently, in the present figure, $\langle P'AR = 120°$, and this means that the arc $P'AR$ is one-third of the circumference of $K$. In precisely the same way, arc $Q'CR$ is also one-third of the circumference, making the minor arc $P'Q'$ the final third. The upshot of this argument is that $P'$, $Q'$, and $R$ determine an *equilateral* triangle inscribed in $K$.

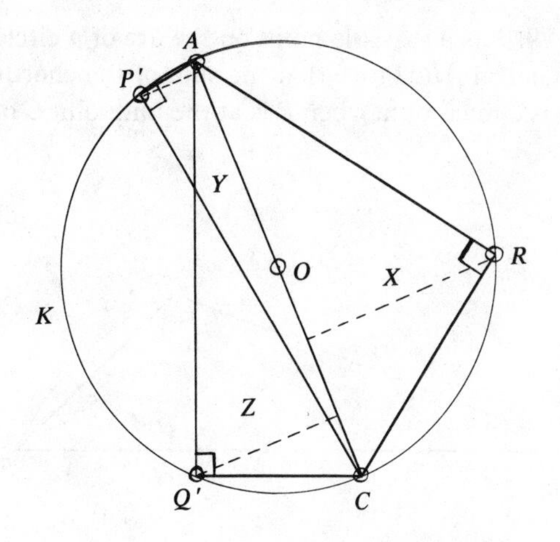


Now then, Mrs. Dijkstra proposes that we suspend equal masses at each of $P'$, $Q'$, and $R$. Since $P'Q'R$ is equilateral, the center of gravity of this system of masses will coincide with the center $O$ of the circle $K$. Thus, if one were to support the point $O$, one would hold up the whole system of masses. A knife-edge placed beneath $AC$, then, in supporting the critical point $O$, would keep the system in equilibrium. Consequently, there would be no tendency for the weighted circle to tip on either side of the edge $AC$. Therefore, the moments about $AC$ determined by the masses at $P'$ and $Q'$ must be balanced precisely by the moment of the mass at $R$, and the desired relation between the altitudes of $X$, $Y$, and $Z$ follows immediately.

What a marvellous solution! (All the more remarkable in view of the fact that Mrs. Dijkstra is an octogenarian.)

## 3. Another Morsel

Mathematical Morsel #9 concerns a simple property that is not difficult to prove. However, it seems to be unusually rich in interesting approaches. It is a great pleasure to pass along two solutions from mathematicians at the University of Technology in Eindhoven, The Netherlands—first, one that is full of ingenuity, by I. van Yzeren, and second, a most elegant one by Karel A. Post.

Morsel #9. $P$ is a variable point on the arc of a circle cut off by a chord $AB$. Prove that the sum of the chords $AP$ and $PB$ is a maximum when $P$ is at the midpoint $C$ of the arc $AB$.

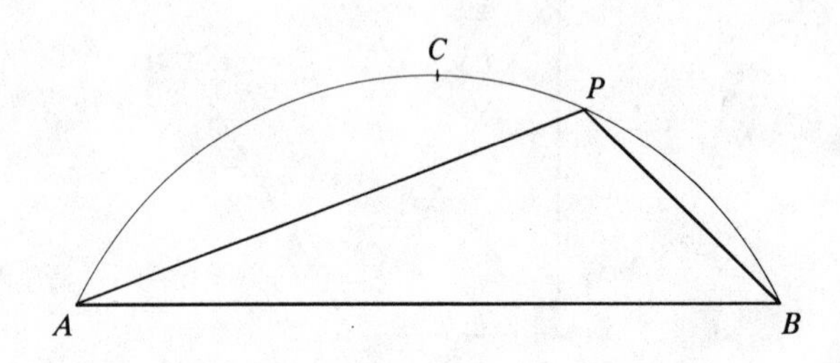


(a) *Van Yzeren's Proof.* In the figure, let $O$ denote the center of the circle and let $OE$ and $OF$ bisect the angles $AOP$ and $POB$. Then $OE$ is the perpendicular bisector of $AP$. Thus we have $PW = (1/2)AP$. But, in sector $EOP$, we have by symmetry that the perpendicular $EX = PW$. Hence $EX = (1/2)AP$, and, similarly, $FY = (1/2)PB$. Consequently we have $(1/2)(AP + PB) = EX + FY$.

Now $EF$ subtends at $O$ an angle of $(1/2) < AOB$, as does $AC$ (and $CB$). This makes $EF = AC$ $(= CB)$.

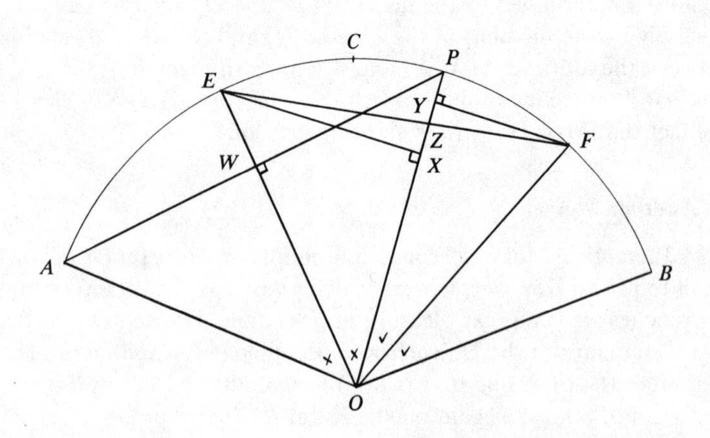

Finally, suppose $EF$ crosses $OP$ at $Z$. Then $EZ$ and $ZF$ are the hypotenuses of right-triangles $EXZ$ and $FYZ$; thus we have

$$\frac{1}{2}(AP + PB) = EX + FY < EZ + ZF = EF$$

$$= AC = \frac{1}{2}(AC + CB),$$

implying the desired $AP + PB < AC + CB$.

(b) *Post's Proof.* Suppose $P$ lies between $C$ and $B$, and that $AP$ crosses $CB$ at $X$. Then the triangles $ACX$ and $BXP$ are equiangular, and therefore similar, with $ACX$ clearly the bigger of the two. Thus, for some $k > 1$, the sides of $ACX$ are $k$ times the corresponding sides of $PXB$:

$$AC = k \cdot PB, \qquad CX = k \cdot PX, \qquad AX = k \cdot BX.$$

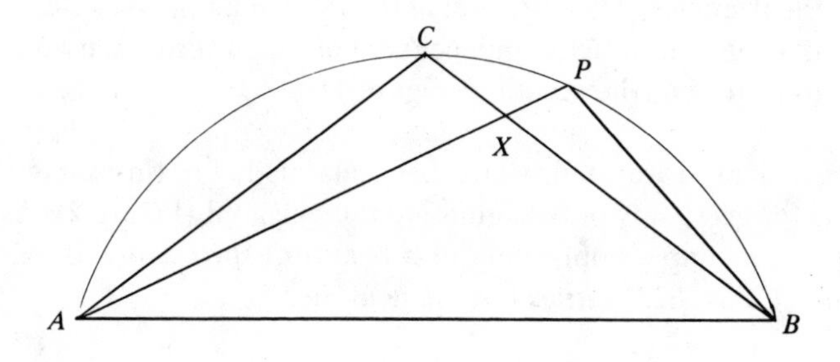


In this case, the vital difference

$$\begin{aligned}(AC + CB) &- (AP + PB)\\ &= (AC + CX + XB) - (AX + XP + PB)\\ &= (AC + CX - AX) - (XP + PB - XB)\\ &= (k \cdot PB + k \cdot PX - k \cdot BX) - (XP + PB - XB)\\ &= (k - 1)(PB + PX - BX),\end{aligned}$$

which, from $k > 1$ and the triangle inequality, is the product of two *positive* factors. Thus this difference is positive, and we have the desired

$$AC + CB > AP + PB.$$

## 4. An Old Japanese Theorem

In Roger Johnson's marvellous old geometry text—*Advanced Euclidean Geometry*, first published in 1929—he reports (on page 193 of the Dover edition, 1960) on the ancient custom by Japanese mathematicians of inscribing their discoveries on tablets which were hung in the temples to the glory of the gods and the honor of the authors. The following gem is known to have been exhibited in this way in the year 1800.

> Let a convex* polygon, which is inscribed in a circle, be triangulated by drawing all the diagonals from one of the vertices, and let the inscribed circle be drawn in each of the triangles. Then the sum of the radii of all these circles is a constant which is independent of which vertex is used to form the triangulation (Figure 1).

A great deal more might have been claimed, for this same sum results for *every* way of triangulating the polygon! (Figure 2). As we shall see, a simple application of a beautiful theorem of L. N. M. Carnot (1753–1823) settles the whole affair.

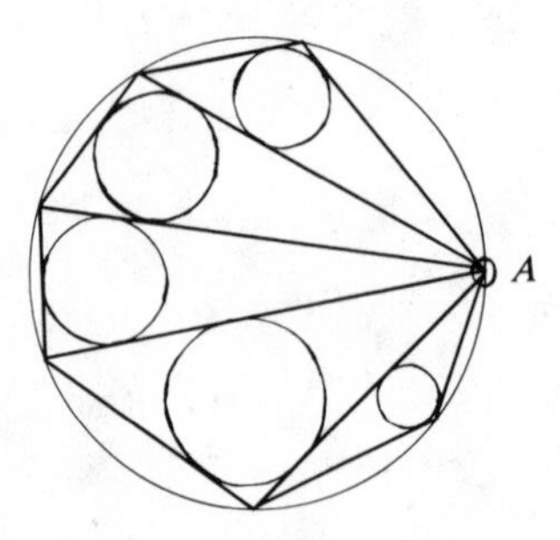


Figure 1

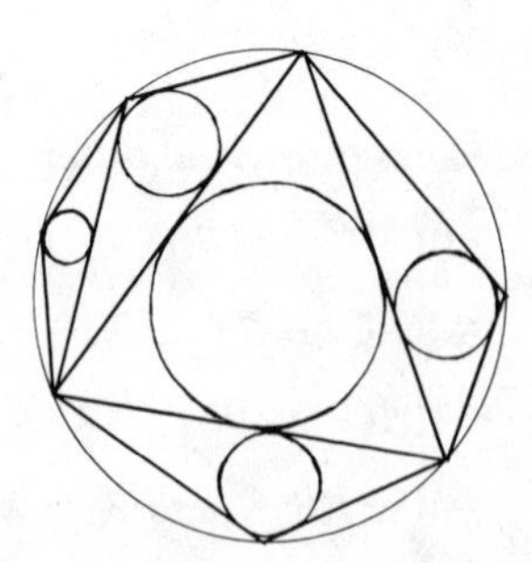

Figure 2

CARNOT'S THEOREM. *In any triangle $A_1A_2A_3$, the sum of the distances (suitably signed) from the circumcenter* $O$ to the sides, is $R + r$, the sum of the circumradius* and the inradius*:*

$$OO_1 + OO_2 + OO_3 = R + r.$$

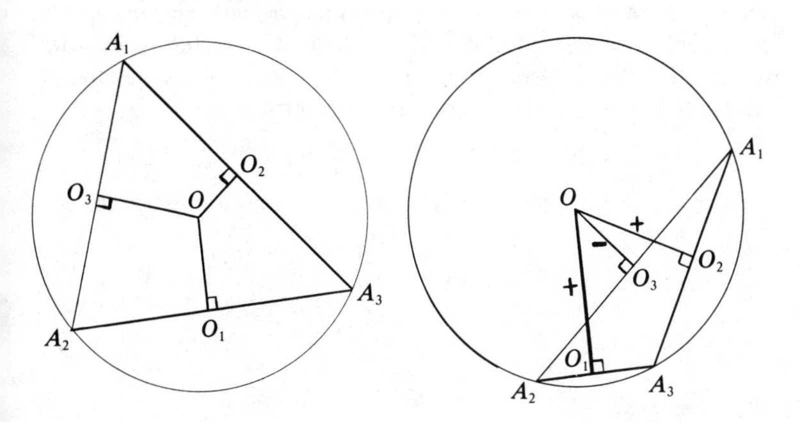


Of course, the center $O$ does not always belong to the triangle itself. In order to cover all cases, there is the following rule for attaching signs to the distances: $OO_i$ is negative if and only if the entire segment $OO_i$ lies outside the triangle (this guarantees that the areas of the 3 triangles $OA_1A_2$, $OA_2A_3$, $OA_3A_1$ always add up to $\Delta A_1A_2A_3$). We will not prove Carnot's theorem here, but the interested reader will find all the clues for a proof on page 190 of Johnson's book (part f).

Let us suppose that the triangles in the triangulation are numbered, and that the inradius of triangle $i$ is $r_i$; moreover, suppose that the distances from the center $O$ (of the large circle) to the sides of triangle $i$ are $OO_1^{(i)}$, $OO_2^{(i)}$, $OO_3^{(i)}$. Then, observing that every triangle in the triangulation has the given circle $O(R)$ (center $O$, radius $R$) as its circumcircle, Carnot's theorem gives

$$r_i + R = OO_1^{(i)} + OO_2^{(i)} + OO_3^{(i)},$$

and the sum in question is given by

$$S = \Sigma\, r_i = \Sigma\,(OO_1^{(i)} + OO_2^{(i)} + OO_3^{(i)} - R)$$
$$= \Sigma\,(OO_1^{(i)} + OO_2^{(i)} + OO_3^{(i)}) - \Sigma\, R.$$

Now, one thing that we know is independent of the triangulation is the *number* of triangles it contains (adding the angles reveals there are $n - 2$ triangles in any triangulation of an $n$-gon). This makes the quantity $\Sigma\, R$ the same for all triangulations. Consequently, in order to establish that the entire sum $S$ is independent of the method of triangulation, we need only show that the term

$$S' = \Sigma\,(OO_1^{(i)} + OO_2^{(i)} + OO_3^{(i)})$$

is constant for all triangulations. But this is almost immediate.

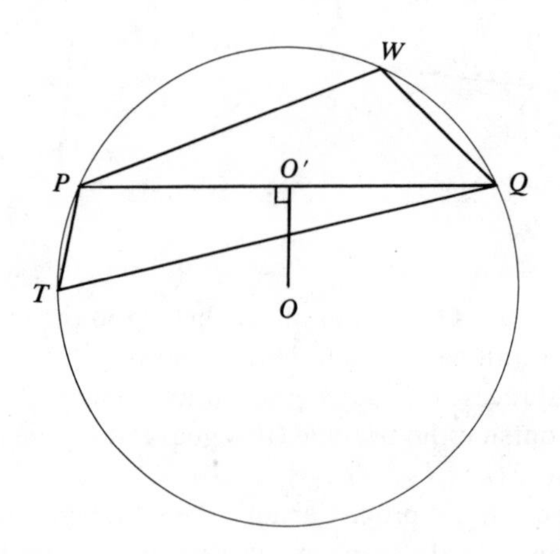


Consider any *diagonal* $PQ$ of the triangulation. Cutting across the polygon as it does, it is a side of *two* triangles in the triangulation, one on each side of it. The perpendicular $OO'$ from $O$, then, necessarily cuts across one of these triangles while remaining entirely outside the other. Thus, in the sum $S'$, the distance $OO'$ occurs twice to no effect, for it occurs once with a plus sign and once with a minus sign. As a result, the value of $S'$ is simply the sum of all the perpendiculars from $O$ to the sides of the given polygon, making it the same for all triangulations.

## 5. Regular Polygons of Unit Side

The *Pi Mu Epsilon Journal* has an outstanding problems section. The following engaging result appeared in Problem 390 of the Spring issue, 1978; the problem was proposed and solved by Robb Koether and David C. Kay, University of Oklahoma, Norman:

> Let $AB\ldots C$ be a regular polygon of unit side. Consider the triangles at $A$ that are determined by $BC$ and the diagonals at $A$. Then, for *each* of these triangles, the length of one of the sides is equal to the *product* of the lengths of the other two sides.

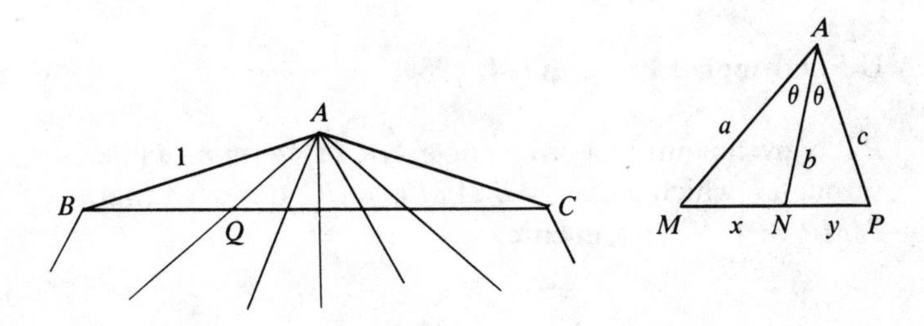


Let $AMN$ and $ANP$ be two adjacent triangles, having sides of lengths $a$, $b$, $c$, $x$, $y$, as marked in the figure. Now, the vertices of a regular polygon lie on a circle, and in this circumcircle the angles $MAN$ and $NAP$ are subtended by equal sides of the polygon. Thus these angles are equal; in fact, all the angles at $A$ are equal. Therefore, $AN$ bisects angle $MAP$, and we can apply the standard theorem "*the bisector of an angle of a triangle divides the opposite side in the same ratio as the sides about the angle*" ([2]):

$$\frac{x}{y} = \frac{a}{c}.$$

Rearranging this, we obtain

$$\frac{a}{x} = \frac{c}{y},$$

and, multiplying by $b$, we get

$$\frac{ab}{x} = \frac{bc}{y}.$$

This declares that the product of the two sides that meet at $A$ gives a constant result (call it $k$) when divided by the side that lies along $BC$.

Suppose $ABQ$ is the triangle having side $AB$. By symmetry, we have $\angle ABQ = \angle BAQ$, making $AQ = BQ$. Thus the constant $k$ is given by

$$k = \frac{AB \cdot AQ}{BQ} = AB = 1,$$

and we have in general that $ab = x$.

## 6. A U.S. Olympiad Problem (#4, 1980)

> $P$ is a given point in a given angle $ABC$. Determine a line through $P$ which crosses $AB$ at $M$ and $BC$ at $N$ such that $1/MP + 1/NP$ is a maximum.

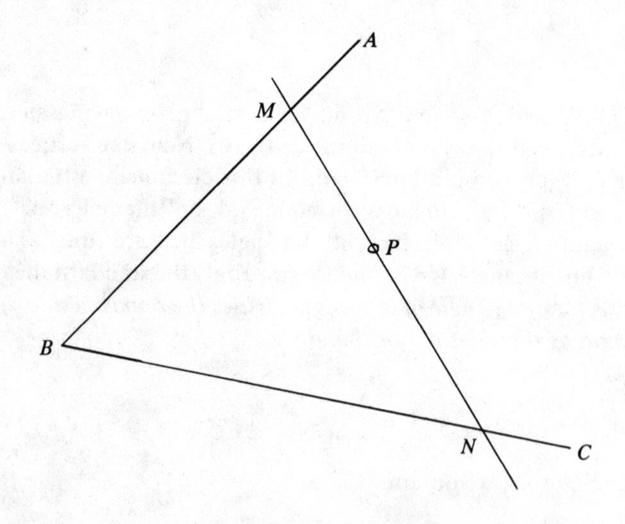


The presence of the inverses of $MP$ and $NP$ set my friend Frank Allaire to wondering whether this might be a job for the method of circular inversion [3] (Frank is now at Lakehead University, Thunder Bay, Ontario). His neat solution follows.

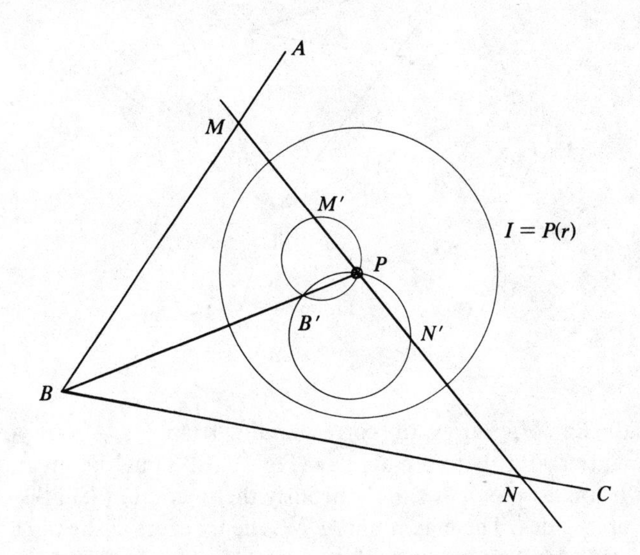


Let the figure be inverted in any circle $I$ having center $P$, say $P(r)$ (Figure 2). The sides $AB$ and $BC$ of the given angle invert into circles through $P$ which intersect a second time at the image $B'$ of $B$. Now, any line $MN$ through the center $P$ inverts into itself. If $M'$ and $N'$ are the images of $M$ and $N$, then we have

$$MP \cdot M'P = r^2 \qquad \text{and} \qquad NP \cdot N'P = r^2,$$

from which

$$\frac{1}{MP} = \frac{1}{r^2} M'P \qquad \text{and} \qquad \frac{1}{NP} = \frac{1}{r^2} N'P.$$

From these inversions, we can turn the fractions "right side up":

$$\frac{1}{MP} + \frac{1}{NP} = \frac{1}{r^2}(M'P + N'P) = \frac{1}{r^2} M'N'.$$

Consequently, a maximum is obtained when the compound chord $M'N'$ is greatest. Fortunately this is easy to determine.

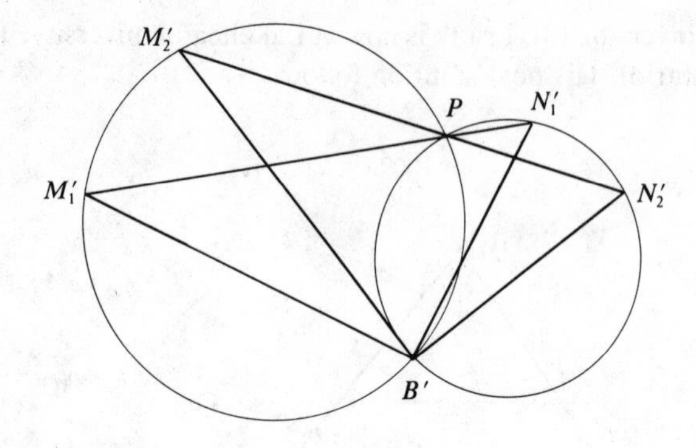


Figure 3

As the line $MN$ varies, the corresponding triangles $M'N'B'$ are all equiangular and therefore similar (Figure 3). Thus the greatest of these triangles provides simultaneously the maximum lengths of all three of its sides. The maximum $M'N'$, then, occurs at the same time as the side $M'B'$ is greatest. But clearly this occurs when $M'B'$ is a diameter, making $\langle M'PB'$ a right angle. The required line $MN$, then, is perpendicular to $BP$.

## 7. Kürschák's Tile

If a regular $n$-gon* $XYZ\ldots$ is inscribed in a unit circle, each side subtends an angle of $2\pi/n$ at the center $O$ and the area $A$ of the polygon is given by

$$A = n(\Delta OXY) = n\left(\frac{1}{2}\cdot 1\cdot 1\cdot \sin\frac{2\pi}{n}\right) = \frac{n}{2}\sin\frac{2\pi}{n}.$$

Now the value of $\sin(2\pi/n)$ is irrational for all $n$ except 4 and 12, which yield $\sin(\pi/2) = 1$ and $\sin(\pi/6) = 1/2$, making the areas of the inscribed square and dodecagon not only rational but integral, with values 2 and 3, respectively. The devoted teacher and tireless

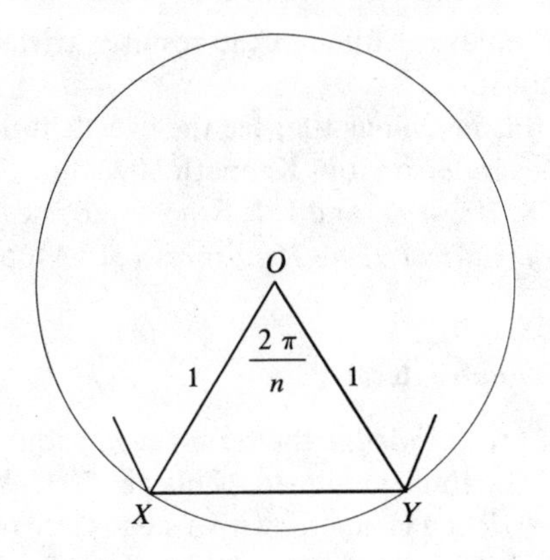


promoter of mathematics J. Kürschák of Hungary (1864–1933) devised the following geometric demonstration for the case of the dodecagon.

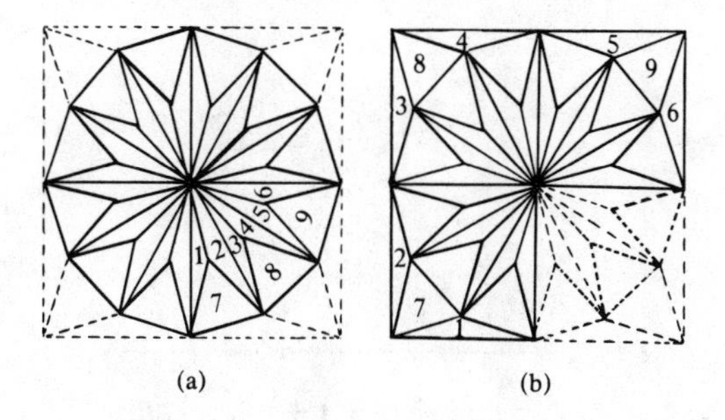


(a) (b)

Let a (2 × 2) square be circumscribed about the circle (not shown) and dodecagon and let the resulting figure be decomposed as shown in figure (a). There are only two kinds of triangle here—(i) equilateral triangles of side equal to the side $s$ of the dodecagon, and (ii) 15°-15°-150° isosceles triangles with equal arms of length $s$ (the easy straightforward verifications of these claims are left to the reader). Then, taking up the nine triangles 1, 2, . . . , 9 in figure (a) from one quarter of the dodecagon, it is clear from figure (b) that the dodecagon can be rearranged to cover exactly three of the unit-square quarters of the large square. Although the result is trivial, what a fabulous way to establish it!

For more on this ingenious tile, see the excellent note on the subject by G. L. Alexanderson and Kenneth Seydel in *The Mathematical Gazette*, 1978, 192–196, and I. J. Schoenberg's collection of popular essays *Mathematical Time Exposures* (MAA), page 7.

## 8. The Equilic Quadrilateral

To the best of my knowledge the term "equilic quadrilateral" was coined by Jack Garfunkel (Queens College, New York), who for many years has enriched geometry with a steady stream of interesting problems. A quadrilateral $ABCD$ is equilic if a pair of opposite sides are equal, $AD = BC$, and are inclined at 60° to each other. Because the opposite sides of many quadrilaterals do not intersect, the latter condition might be more conveniently stated in the form $\langle A + \langle B = 120^\circ$.

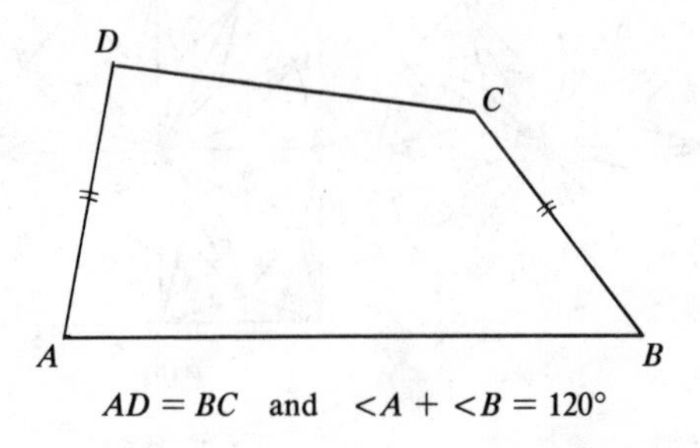


$AD = BC$ and $<A + <B = 120^\circ$

A very readable introduction to this configuration is given by Professor Garfunkel in the *Pi Mu Epsilon Journal*, fall 1981, 317–329. Among the properties discussed there are the following engaging results.

1. *If $ABCD$ is an equilic quadrilateral with $AD = BC$ and $\langle A + \langle B = 120^\circ$, then the midpoints $P$, $Q$, and $R$ of the diagonals and the side $CD$ always determine an equilateral triangle.*

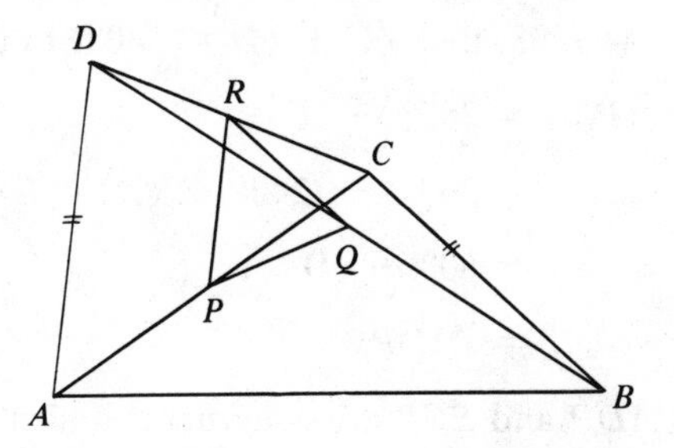


Since $P$ and $R$ are the midpoints of $AC$ and $DC$, $PR$ is parallel to $AD$ and of length $(1/2)AD$. Similarly, $QR$ is parallel to $BC$ and of length $(1/2)BC$. Because $AD = BC$, then $PR = QR$. Also, because of the parallels, $PR$ and $QR$ are inclined at the same angle as $AD$ and $BC$ $(60°)$. This makes $\Delta PQR$ equilateral.

2. *If equilateral triangle PCD is drawn outwardly on CD, then $\Delta PAB$ is also equilateral.*

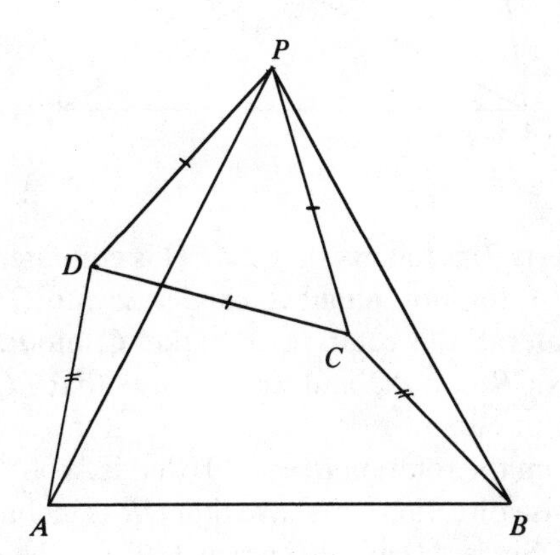

Since $\langle A + \langle B = 120°$, then $\langle C + \langle D = 240°$. In this case,

$$\begin{aligned}\langle PCB &= 360° - \langle C - 60° \\ &= 300\,° - (240° - \langle D) \\ &= 60° + \langle D \\ &= \langle ADP.\end{aligned}$$

Thus triangles $ADP$ and $BCP$ are congruent, and it follows easily that, in $\Delta ABP$, the angle between the equal $AP$ and $BP$ is $60°$.

3. *If equilateral triangles are drawn on $AC$, $DC$ and $DB$, away from $AB$, then the three new vertices $P$, $Q$, and $R$ are collinear.*

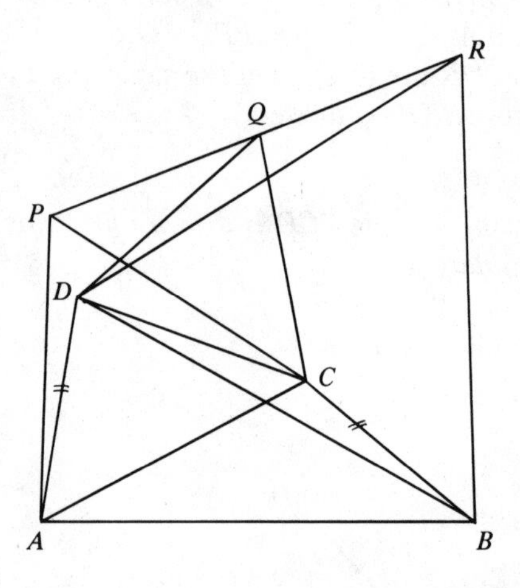


By our property 2, it follows that $\Delta ABQ$ is equilateral. Thus a $60°$ counterclockwise rotation about $A$ carries $B$ into $Q$. But, because $\Delta PAC$ is equilateral, this rotation also takes $C$ into $P$. Thus this $60°$ rotation carries $CB$ into $PQ$ and we conclude that $PQ$ is inclined to $CB$ at an angle of $60°$.

Similarly, from the rotation $B(-60°)$ (that is, a $60°$ rotation about $B$ in the clockwise direction), we have that $QR$ is inclined to $AD$ at an angle of $-60°$. Since $AD$ and $BC$ are at $120°$ (or the supplementary

60°, depending on the viewpoint you take), turning one through 60° and the other through −60°, in the appropriate directions, brings them to the same direction, and we have that $PQ$ and $QR$ line up in that common direction.

**Exercises**

$ABCD$ is equilic with $AD = BC$ and $\langle A + \langle B = 120°$

1. Equilateral triangles drawn outwardly on $AD$, $AB$, and $BC$ yield three new vertices which determine an equilateral triangle.

2. Equilateral triangles drawn outwardly on $AD$ and $DC$, and inwardly on $BC$, yield three new vertices that determine an equilateral triangle.

3. Equilateral triangles drawn inwardly on $AD$, $CD$ and $BC$ yield three new vertices which determine an equilateral triangle.

4. Reflect $ABCD$ in $AB$ to get hexagon $ADCBC'D'$. Then equilateral triangles drawn outwardly on either set of alternate sides yield three new vertices which determine an equilateral triangle.

**References**

1. R. Honsberger, Mathematical Morsels, Dolciani Mathematical Expositions, vol. 3, Mathematical Association of America, 1978.
2. R. A. Johnson, Advanced Euclidean Geometry, Dover, Mineola, N.Y., 1960, chapter 4.
3. H. S. M. Coxeter and S. L. Greitzer, Geometry Revisited, New Mathematical Library Series, vol. 19, Mathematical Association of America, 1967, chapter 5.

CHAPTER 3

# TWO PROBLEMS IN COMBINATORIAL GEOMETRY

Combinatorial geometry ([**1**], [**2**]) is concerned with counting things that arise in geometric settings. It never occurred to the Greeks to ask *how many* points, lines, triangles, etc., there might be in their figures. This is a thoroughly modern subject that has shot ahead in the last 40 years. In this section I would like to show you two nuggets from this mathematical gold mine.

## 1. An Elusive Set of Points

In the first problem we shall be concerned with the *distances* that are determined by the pairs of points of a set of points in the plane. On the surface of things, one might wonder what there would be to say about such a collection of distances. Presumably they might all be the same or, perhaps, all different. In fact, however, this is generally not so. Would you believe that no matter where $n$ points ($n \geq 3$) might be chosen in the plane

(i) the number of *different* distances produced must be at least $\sqrt{n - 3/4} - 1/2$,

(ii) the smallest distance produced cannot occur more often than $3n - 6$ times,

(iii) the greatest distance produced cannot occur more often than $n$ times,

(iv) no distance produced can occur more than $n^{3/2}/\sqrt{2} + n/4$ times?

The elementary proofs of these intriguing facts are given in my book *Mathematical Gems II* [**3**]. They were discovered in 1946 by one of those incredible Hungarians—Paul Erdös—who, with colleagues like Paul Turán and Tibor Gallai, did so much to bring combinatorial geometry into prominence.

Our first problem comes from the first-class British periodical *Mathematical Spectrum* [**4**]. It asks that we

> determine, for any given positive integer $n$, a set of points in the plane, containing any finite number of points you please, having the demanding property that *every point in the set be a distance of 1 unit from exactly n other points in the set.*

What a condition—requiring this of *every* point in the set! The following beautiful solution is due to David Beal (Winchester College, England).

He proceeds by induction. Clearly the endpoints of a unit segment constitute an acceptable set when $n = 1$, and the vertices of an equilateral triangle of unit side dispose of the case $n = 2$. Suppose, then, that a set $S$ is known to be acceptable for the case $n = k - 1$, and that we address ourselves to the case $n = k$.

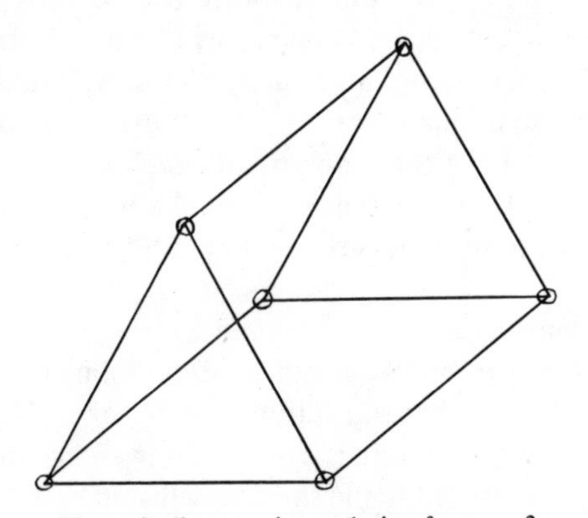

The unit distances in a solution for $n = 3$

Suppose $S$ consists of the $t$ points $p_1, p_2, \ldots, p_t$. Now, let $S$ be translated* a distance of 1 unit, in a direction $d$ to be determined, to take up the positions $q_1, q_2, \ldots, q_t$. If we avoid sliding the set in the direction of any segment $p_i p_j$, then no point $p_i$ will be carried to a position on top of a point $p_j$, implying that the images $q$ will be $t$ entirely new points. With one additional stipulation on the direction $d$, the resulting set $R$, consisting of these $2t$ $p$'s and $q$'s, will be the set we are looking for.

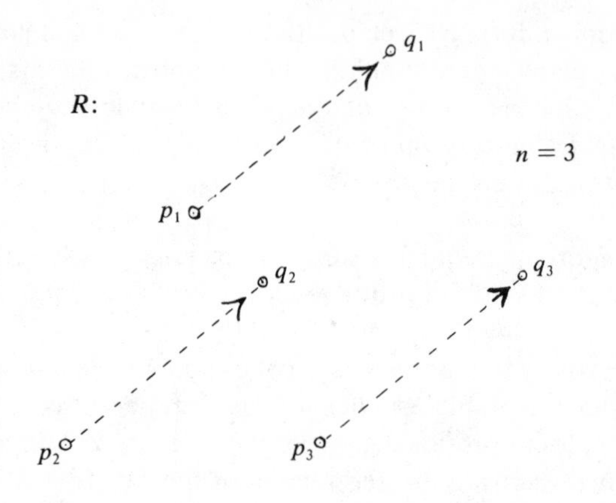


Because each point $p_i$ is a unit distance from exactly $k - 1$ other points $p$ (say $p_a, p_b, \ldots$), its image $q_i$ will be a unit distance from exactly $k - 1$ other points $q$ (namely the corresponding $q_a, q_b, \ldots$). Since the translation is through a unit distance, then $p_i q_i = 1$ for every $i$, and we have that each $p_i$ and each $q_i$ is a unit distance from a minimum of $k$ other points of $R$. We would like to arrange things so that this minimum is the actual total for every point in the set.

First, let us consider a point $p_i$. Distances from $p_i$ are either $p_i p_j$'s or $p_i q_j$'s. But the $p_i p_j$'s are beyond manipulation, for they are already fixed in $S$ before the translation is undertaken. Thus the only thing we have to worry about is creating unwanted unit distances among the $p_i q_j$. We are led to the same conclusion by a consideration of the distances from a point $q_i$. Hence the question is "can we avoid making any $p_i q_j = 1$ for $i \neq j$?" The answer is yes; in fact, it is very easy.

We simply need to avoid moving a point $p_j$ to a position on the unit circle with center $p_i$. For a given $p_i$, this eliminates at most two directions for each point $p_j$. Taking into account all the pairs of points $(p_i, p_j)$, and the inadvisable directions $p_i p_j$, noted earlier, a great many directions might be ruled out. However, for any $n$, the number of unsafe directions will be finite, leaving any number of satisfactory choices for $d$, and the conclusion follows by induction.

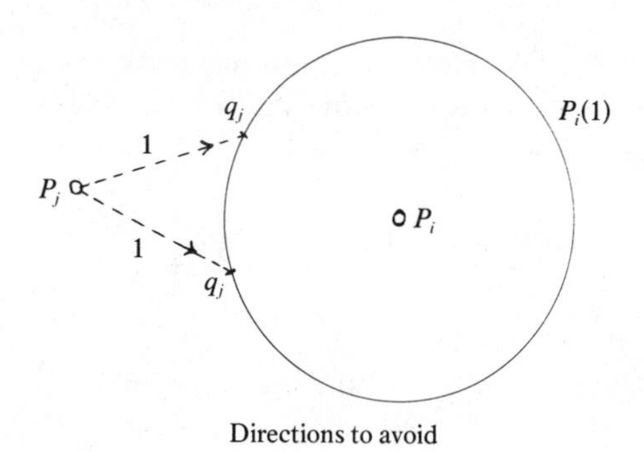


Directions to avoid

## 2. A Problem about Triangles

Our second problem was very kindly brought to my attention by the outstanding number theorist George Andrews (Pennsylvania State University).

> How many different triangles are there which have integral sides and perimeter $n$?

A complete specification of this number $T(n)$ is given in a paper by J. H. Jordan, Ray Walch, and R. J. Wisner [**5**]. In a brief note [**6**] George Andrews gives a beautiful solution which exploits the rather natural connection between $T(n)$ and $p_3(n)$ and $p_2(n)$, the number of ways of partitioning an integer $n$ into 3 and 2 parts, respectively.

*Andrews' Solution.* A partition of $n$ into 3 parts, $a + b + c = n$, generally defines a triangle counted by $T(n)$. The only way a partition will not do so is by failing to satisfy one of the triangle inequalities

$$a + b > c, \qquad b + c > a, \qquad c + a > b.$$

But this can happen only when the sum of the two smaller parts, say $b + c$, fails to exceed the largest part $a$: i.e., when $b + c \leq a$. In this case, $b$ and $c$ would add up to some integer $j \leq (1/2)n$, which, because only integers are involved, is equivalent to

$$b + c = j \leq \left[\frac{1}{2}n\right],$$

where $[x]$ is used to denote the greatest integer $\leq x$.

Conversely, suppose $j$ is a positive integer $\leq [(1/2)n]$. Then each of the $p_2(j)$ partitions of $j$ into 2 parts,

$$b + c = j \qquad \left(\leq \left[\frac{1}{2}n\right]\right),$$

gives

$$b + c + j \leq n, \qquad \text{and} \qquad b + c \leq n - j.$$

Letting $n - j = a$, we have $b + c \leq a$, where

$$a + b + c = n - j + b + c = n.$$

That is to say, each of the $p_2(j)$ partitions $(b, c)$ corresponds to a partition $(a, b, c)$ of $n$ that, in view of $b + c \leq a$, fails to generate a triangle. Thus there is a 1-1 correspondence between the failing partitions $(a, b, c)$ of $n$ and the partitions into 2 parts of the integers $j$ in the range 1 to $[(1/2)n]$. Subtracting from the $p_3(n)$ possible partitions the total number of failures, we obtain

$$T(n) = p_3(n) - \sum_{1 \leq j \leq [(1/2)n]} p_2(j).$$

Now, by simply listing all the partitions, we see that $p_2(j)$ is always just $[(1/2)j]$:

for $j = 2k + 1$, the partitions are $(1, 2k)$, $(2, 2k - 1)$, $\ldots$, $(k, k + 1)$;

for $j = 2k$, the partitions are $(1, 2k - 1)$, $(2, 2k - 2)$, $\ldots$, $(k, k)$;

thus $p_2(j) = k = [(1/2)j]$.

Using this result, it is an exercise in mathematical induction to establish that the sum in question is given by $[n/4][(n+2)/4]$, yielding

$$T(n) = p_3(n) - \left[\frac{n}{4}\right]\left[\frac{n+2}{4}\right].$$

But, as we shall see, $p_3(n)$ is given by the formula $p_3(n) = \{n^2/12\}$, where $\{x\}$ denotes the integer *nearest* $x$ (since no square is ever halfway between two multiples of 12, this is never ambiguous). Finally, then, we have the pretty result

$$T(n) = \left\{\frac{n^2}{12}\right\} - \left[\frac{n}{4}\right]\left[\frac{n+2}{4}\right].$$

*The Formula for* $p_3(n)$. Let us prove this formula for $p_3(n)$. In order to do this, we introduce a simple geometric representation of a partition which is known as a Ferrers graph. In a Ferrers graph $G$, each part $r$ in the partition is represented by a row of $r$ equally spaced dots. In preparing a partition for representation by a Ferrers graph, it needs to be written in nonincreasing order. Accordingly, the rows in a Ferrers graph, which are lined up one under the other from the left so that the dots fall into columns, also occur in nonincreasing order. For example, $10 = 4 + 2 + 2 + 1 + 1$ yields the Ferrers graph:

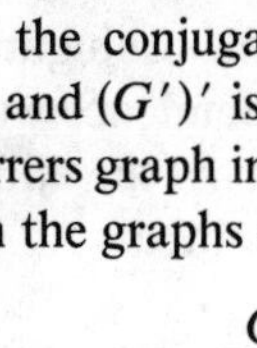

```
. . . .
. .
. .
.
.
```

The Ferrers graph $G'$, which is obtained from $G$ by interchanging its rows and columns, is called the conjugate of $G$. Clearly a graph $G$ has exactly one conjugate $G'$ and $(G')'$ is simply $G$, itself. Thus, if the conjugate is taken of each Ferrers graph in a set $S$, the set of conjugates is in 1–1 correspondence with the graphs of $S$.

```
G:          G':
. . . .     . . . . .
. .         . . .
. .         .
.           .
.
```

Let us consider, then, the collection of Ferrers graphs of the partitions that are counted by $p_3(n)$. Each of these graphs $G$ will have exactly 3 rows, which means that each conjugate $G'$ will have exactly 3 dots in its first row and no more than 3 dots in any other row. As such, each $G'$ represents a partition of $n$ in which every part

| $G$: | $G'$: |
| --- | --- |
| . . . . . | . . . |
| . . . | . . . |
| . . | . . |
| | . |
| | . |

is either a 1, 2, or 3, and at least one part must be a 3. Since $p_3(n)$ counts *all* partitions containing exactly 3 parts, it is not difficult to see that the set of conjugates $\{G'\}$ must represent *all* partitions of this kind (i.e., with parts that are 1's, 2's, or 3's, with at least one 3). In fact, the whole purpose of these Ferrers graphs and their conjugates is simply to provide us with a nice way of seeing that there exists a 1-1 correspondence between the partitions counted by $p_3(n)$ and the partitions of $n$ into 1's, 2's, and 3's, where at least one 3 must occur. We have, then, that the *number* $p_3(n)$ is the same as the number of partitions in this latter class.

Now, if this obligatory 3 is removed from a conjugate partition, we obtain a partition of $n - 3$ into 1's, 2's, and 3's which has no additional qualifying condition on its composition (it may or may not have any 3's left). If this is done to each conjugate partition of $n$, then, a set of partitions of $n - 3$ is obtained which is clearly in 1-1 correspondence with the set of conjugate partitions of $n$, and, therefore, is also in a 1-1 correspondence with the partitions counted by $p_3(n)$. Letting $p(\text{“}A\text{”}, m)$ denote the number of partitions of $m$ which have parts from the set of numbers $A$, we have that

$$p_3(n) = p(\text{“}\{1, 2, 3\}\text{”}, n - 3).$$

At last we have arrived at an expression for $p_3(n)$ that we are able to handle.

In order to calculate this quantity we turn to one of the premier tools of the combinatorialist—generating functions [7]. Consider the product

$$f(x) = (1 + x + x^2 + x^3 + \cdots)(1 + x^2 + x^4 + x^6 + \cdots)$$
$$(1 + x^3 + x^6 + x^9 + \cdots).$$

In multiplying these series together, one of the terms in $x^{16}$, for example, is obtained by taking

$x^3$ from the first factor,

$x^4$ from the second, and

$x^9$ from the third.

This displays the exponent 16 in the form

$$16 = 3 + 4 + 9,$$

which we may construe to be

$$16 = 3(1) + 2(2) + 3(3)$$
$$= 1 + 1 + 1 + 2 + 2 + 3 + 3 + 3,$$

corresponding to a partition of 16 in which only 1's, 2's, and 3's occur (the number of 1's comes from the first factor in $f(x)$, and so on). Conversely, every such partition of 16 can be used as a prescription for selecting terms from the three factors of $f(x)$, based on the number of 1's, 2's, and 3's that are called for, that will generate a term in $x^{16}$. Consequently, the total coefficient of $x^{16}$ in $f(x)$ is just $p(\text{"}\{1, 2, 3\}\text{"}, 16)$, and in general, the desired $p(\text{"}\{1, 2, 3\}\text{"}, n - 3)$ is the coefficient of $x^{n-3}$ in $f(x)$. Fortunately we can determine this coefficient by elementary methods.

First of all, observe that the binomial theorem gives

$$(1 - x^k)^{-1} = 1 + x^k + x^{2k} + x^{3k} + \cdots,$$

making

$$f(x) = (1 - x)^{-1}(1 - x^2)^{-1}(1 - x^3)^{-1}$$
$$= \frac{1}{(1 - x)(1 - x^2)(1 - x^3)}.$$

Now, by resolving $f(x)$ into its partial fractions, we obtain

$$f(x) = \frac{1/6}{(1-x)^3} + \frac{1/4}{(1-x)^2} + \frac{1/4}{1-x^2} + \frac{1/3}{1-x^3}$$

$$= \frac{1}{6}(1-x)^{-3} + \frac{1}{4}(1-x)^{-2} + \frac{1}{4}(1-x^2)^{-1} + \frac{1}{3}(1-x^3)^{-1}.$$

Therefore, the desired coefficient of $x^{n-3}$ is the sum of the coefficients of $x^{n-3}$ that are obtained from these four parts. But these may be extracted by straightforward applications of the binomial theorem. From the first part we get

$$\frac{1}{6}\cdot\frac{(-3)(-4)\cdots[-3-(n-3)+1]}{(n-3)!}(-1)^{n-3}$$
$$= \frac{1}{6}\cdot\frac{3\cdot4\cdots(n-1)}{(n-3)!} = \frac{1}{6}\cdot\frac{(n-2)(n-1)}{2},$$

and from the second part we similarly obtain $(1/4)(n-2)$.
In the third part, only even powers of $x$ occur, and we obtain the coefficient 0 if $n-3$ is odd, and $(1/4)(1) = 1/4$ if $n-3$ is even. We may express this by saying that the coefficient is $(1/4)k$, where $k$ is either 0 or 1. Similarly, in the final part, the coefficient is $(1/3)t$, where $t$ is either 0 or 1. In these terms, then, we have

$$p_3(n) = \frac{1}{6}\cdot\frac{(n-1)(n-2)}{2} + \frac{1}{4}(n-2) + \frac{1}{4}k + \frac{1}{3}t$$

$$= \frac{n^2 - 4 + 3k + 4t}{12}.$$

Now the most that $3k + 4t$ can be is 7 and the least is 0. Therefore, we have

$$\frac{n^2-4}{12} \le p_3(n) \le \frac{n^2+3}{12},$$

that is,

$$\frac{n^2}{12} - \frac{1}{3} \le p_3(n) \le \frac{n^2}{12} + \frac{1}{4}.$$

Thus the integer $p_3(n)$ does not differ from $n^2/12$ by more than $1/3$, making it the integer nearest $n^2/12$, as claimed.

*Another Approach to* $T(n)$. Finally, let us close with a most elegant solution of this problem, which is based on the fact that

$$T(2n) = p_3(n), \tag{1}$$

an insight that was made independently by N. J. Fine and P. Pacitti of Pennsylvania State University. Combined with the property

$$T(2n - 3) = T(2n), \tag{2}$$

the formula for $p_3(n)$ gives another complete solution. Again, I am indebted to George Andrews for this approach.

The key result $T(2n) = p_3(n)$ is established directly by displaying a 1-1 correspondence between the triangles counted by $T(2n)$ and the partitions of $p_3(n)$. Suppose that $(a, b, c)$ is a triangle counted by $T(2n)$. In this case, $a + b + c = 2n$, and because each side of a triangle is less than one-half the perimeter, we have each of $a, b, c < n$. Consequently, each of the integers $n - a, n - b, n - c$ is positive, and $(n - a, n - b, n - c)$ is a partition counted by $p_3(n)$:

$$n - a + n - b + n - c = 3n - (a + b + c) = 3n - 2n = n.$$

Conversely, if $(p, q, r)$ is a partition counted by $p_3(n)$, we have $p + q + r = n$, and that each of $p, q, r$ is less than $n$. Then the 3 positive integers $n - p, n - q, n - r$ add to $2n$ and the sum of any two exceeds the third, for example, $n - p + n - q = 2n - (p + q) > 2n - n = n > n - r$. Therefore, $(n - p, n - q, n - r)$ is a triangle counted by $T(2n)$, and we have

$$T(2n) = p_3(n).$$

Now let us verify property (2). If $(a, b, c)$ is a triangle counted by $T(2n - 3)$, it is easy to see that $(a + 1, b + 1, c + 1)$ is a triangle counted by $T(2n)$: from $a + b + c = 2n - 3$, we have $a + 1 + b + 1 + c + 1 = 2n$, and, from the known $b + c > a$, etc., we have $b + 1 + c + 1 > a + 2 > a + 1$, etc., satisfying the triangle inequalities. Hence $T(2n) \geq T(2n - 3)$.

We shall obtain the desired $T(2n) = T(2n - 3)$ by showing that $T(2n - 3) \geq T(2n)$. To this end, suppose $(a, b, c)$ is a triangle counted by $T(2n)$. First we will show that none of $a, b, c$ can be unity. Suppose to the contrary, for example, that $c = 1$. By the triangle inequality, we would then have that the integer

$$|a - b| < c = 1,$$

which would leave no option but $a = b$. In this case, however, the total perimeter would be

$$a + b + c = 2a + 1,$$

an odd number, not the $2n$ it is supposed to be. Thus each of $a, b, c$ must exceed unity.

Finally, we complete the argument by showing that the triple of positive integers $(a - 1, b - 1, c - 1)$ determines a triangle that is counted by $T(2n - 3)$. Because $(a, b, c)$ is counted by $T(2n)$, it is clear that $a - 1 + b - 1 + c - 1 = 2n - 3$; furthermore, the known triangle inequality

$$a + b > c,$$

for example, yields

$$a - 1 + b - 1 > c - 1 - 1,$$

or

$$a - 1 + b - 1 \geq c - 1;$$

now, if equality were to hold here, then the total perimeter of the triangle $(a - 1, b - 1, c - 1)$ would be $2(c - 1)$, an even number instead of the odd $2n - 3$. Thus the triangle inequalities are satisfied by $(a - 1, b - 1, c - 1)$, and we have the desired $T(2n - 3) \geq T(2n)$.

Consequently, we have

$$T(2n - 3) = T(2n) = p_3(n) = \left\{\frac{n^2}{12}\right\}.$$

This gives

$$T(2n) = \left\{\frac{n^2}{12}\right\} = \left\{\frac{(2n)^2}{48}\right\}$$

and

$$T(2n-3) = \left\{\frac{[(2n-3)+3]^2}{48}\right\}.$$

That is to say,

$$\text{if } n \text{ is even, then } T(n) = \left\{\frac{n^2}{48}\right\},$$

$$\text{if } n \text{ is odd, then } T(n) = \left\{\frac{(n+3)^2}{48}\right\}.$$

**Exercise**

If $a$, $b$, $c$ are positive integers such that $a^2 + b^2 = c^2$ (that is, if $(a, b, c)$ is a Pythagorean triple), prove that

$$p_3(a) + p_3(b) = p_3(c).$$

(Proposed by Jack Garfunkel, Queen's College, New York, in *Pi Mu Epsilon Journal*, 1981, page 31).

**References**

1. H. Hadwiger, H. Debrunner, and V. Klee, Combinatorial Geometry in the Plane, Holt, Rinehart and Winston, New York, 1964.
2. I. M. Yaglom and V. G. Boltyanski, Convex Figures, Holt, Rinehart and Winston, New York, 1961.
3. R. Honsberger, Mathematical Gems II, vol. 2, Dolciani Mathematical Expositions, MAA, 1973.
4. Mathematical Spectrum, 6 (2) (1973/74); also the sequel in 7 (2) (1974/75) 69–70.
5. J. H. Jordan, R. Walch, and R. J. Wisner, Triangles with integer sides, Amer. Math. Monthly, 86 (1979) 686–689.
6. G. Andrews, A note on partitions and triangles with integer sides, Amer. Math. Monthly, 86 (1979) 477.
7. A. Tucker, Applied Combinatorics, John Wiley & Sons, New York, 1980.

CHAPTER **4**

# SHEEP FLEECING WITH WALTER FUNKENBUSCH

At the 1982 Ontario Mathematics Meeting at Lakehead University in Thunder Bay, Walter Funkenbusch (Michigan Technological University) gave a marvellous talk on some great barroom wagers. Among the items he presented were the following nefarious schemes.

Picture Walter ($W$), complete with eyeshade, string tie and arm bands, sitting at a Las Vegas gaming table, greeting his first Patsy ($P$) of the evening. To begin the action, $W$ produces a set of 3 dice—one red, one white, and one blue. Good-natured to a fault, $W$ lets you ($P$) choose any die you like, after which he takes one of the remaining two, and you both roll for the higher number. This has to sound pretty good because, if you find that Walter is taking the blue one (or whatever color) a lot, you can appropriate any advantage by taking it yourself.

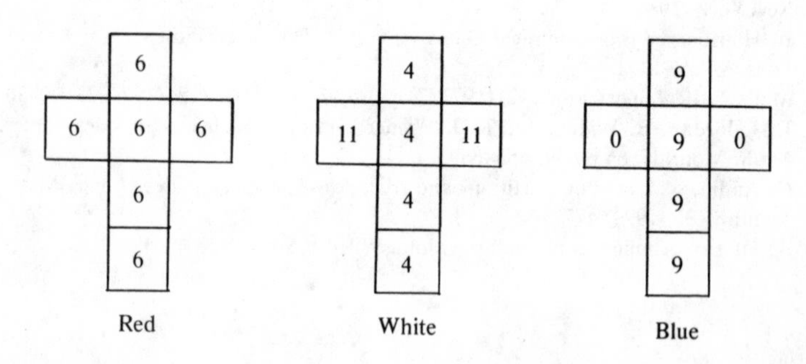


When you begin to show signs of getting tired of losing at this game, Walter confesses that, because he has taken a particular liking to you, he is going to give you an unprecedented opportunity to recover your losses, a move that would surely cost him his job if the management found out about it. You can pick any of the three dice for yourself *and even pick one for Walter, too*. All Walter asks is the right to request, if he is so disposed, that both of you roll *two* copies of your designated die instead of just one. Surely this can't help Walter's chances: if you have given him the poorer of the selected dice, then rolling two of each is only going to make it twice as hard for him to win. How can you lose?

After another series of withering defeats, Walter is simply incredulous at your inexplicable run of bad luck. Perhaps things will take a turn for the better if you play a slightly different game, whereupon he brings out 6 additional dice, 2 each of $X$, $Y$, and $Z$ below. $X$ is colored red and white, indicating that it is to be used only when the two dice you select (for yourself and Walter) are the red and white ones; similarly, $Y$ is white and blue, and $Z$ is blue and red. There are two each of $X$, $Y$, and $Z$, so that, when the $X$'s are used, you can each have one, and so on. On the side, Walter whispers that this is only camouflage, so he can continue to offer you highly favorable odds without the management catching on. Obviously, rolling identical second dice can't help anybody.

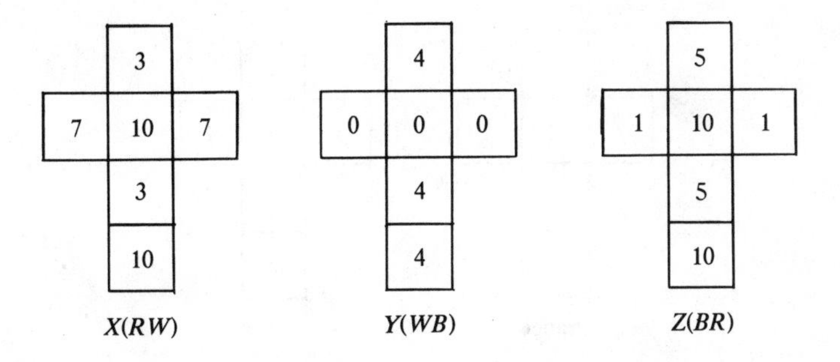


You still get to pick the dice for both players. Then the following three possibilities are considered:

(i) rolling the selected dice as they stand,
(ii) each player rolling 2 copies of his selected die,
(iii) each player rolling the same additional die (the appropriate member of $X$, $Y$, $Z$), along with his selected one.

Walter doesn't want to make any real decisions himself; you can handle the details as you wish. All Walter asks is that he be allowed to rule on one trifling matter. You begin by selecting a die for each player and then Walter either accepts that choice and you proceed to roll as in (i), or he elects to have you take *your choice* between possibilities (ii) and (iii). What could be fairer?

One thing you have to say about Walter is that he is a man of great compassion. As your losses continue to mount, he is obviously deeply moved. When you signed over the deed to your house, he was inconsolable, and could hardly be persuaded to accept it. But that was the limit; he simply had to find some way for you to get even.

Suddenly Walter remembered another game that he was sure couldn't fail to do the trick. Clearing away all the old dice, he produced 3 copies of each of the following orange and green dice. In the new game, one player is to get his choice of color and the other is to decide on how many of each should be rolled, each player rolling the same number of dice (one rolling copies of the orange die, the other the same number of green ones). You are permitted to choose the matter you would like to settle (either the colors or the number), and Walter will decide on the remaining item.

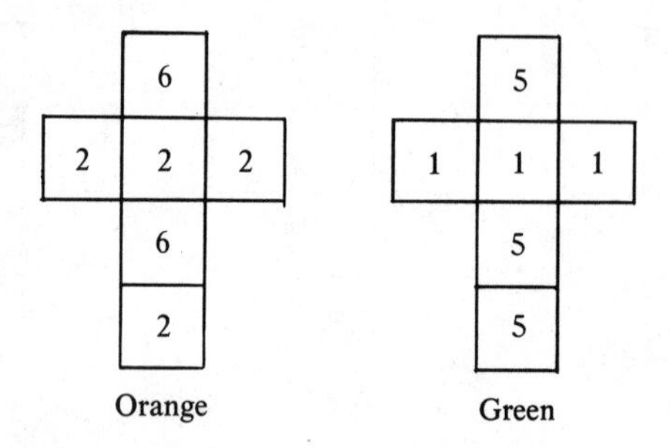


Again noticing that you seem to be losing quite often, Walter suggests two changes that will guarantee a reversal in your fortunes. He proposes to reduce the odds against you by eliminating one die of each color (you did lose a lot when there were 3 of each color) and, more importantly, he will let you have control by *going first himself.* What more could any kind, thoughtful person do?

It soon became evident that your run of bad luck was of epic proportions. In a last-ditch attempt to save you from the depths of despair, Walter pulled out all the stops with a proposition that nobody in his right mind could refuse. He put away all the old props and brought out the following 3 dice:

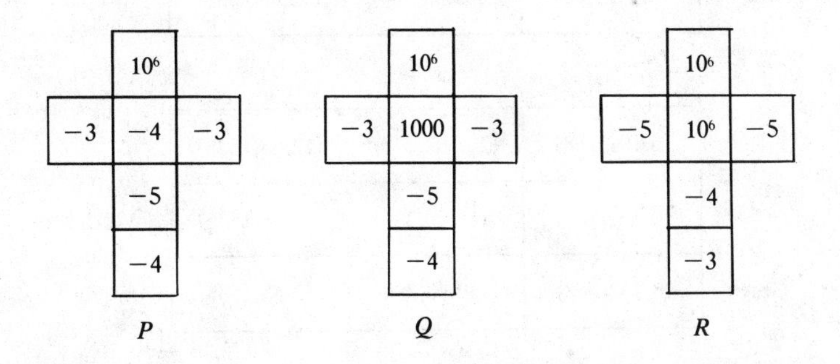


Now here is the ultimate offer. Of the 3 dice $P$, $Q$, $R$, you can roll *two* of them (to Walter's one); as usual, the higher total wins, *and you can choose whichever two you like* (Walter will simply roll the one you leave, and it's pretty obvious which one that will be). Good grief, what a noble character!

When it was discovered that you were absolutely dead broke, Walter was observed to beam with pride and satisfaction. And why not; Walter is a great artist.

Walter Funkenbusch and Donald Saari (Northwestern University) have published a refreshingly readable paper on this subject in the journal *Congressus Numerantium* [**1**]. Let's look now at an amazing set of "triply-nested nontransitive cyclic" dice which are described in their paper.

The set contains $3^3 = 27$ dice in all, 9 of each size—large, medium, and small; 9 of each color—red, yellow, and green, each size containing 3 of each color; and, finally, the 3 of a given size and color distinguished by some other criterion, say a letter $A$, $B$, or $C$. Each die contains only 3 numbers on its faces, each occurring on 2 faces; we use $(a, b, c)$ to denote the die having two faces numbered $a$, two numbered $b$, and two numbered $c$. Since each face is equally likely to

| | A | B | C | |
|---|---|---|---|---|
| Red | (6,192,768) | (24,102,378) | (12,48,1536) | |
| Yellow | (6,192,768) | (18,97,389) | (12,48,1536) | Large |
| Green | (6,192,768) | (30,90,384) | (12,48,1536) | |

| | A | B | C | |
|---|---|---|---|---|
| Red | (6,192,768) | (24,102,378) | (12,48,1536) | |
| Yellow | (6,192,768) | (17,96,391) | (12,48,1536) | Medium |
| Green | (6,192,768) | (30,90,384) | (12,48,1536) | |

| | A | B | C | |
|---|---|---|---|---|
| Red | (6,192,768) | (24,102,378) | (12,48,1536) | |
| Yellow | (6,192,768) | (19,95,390) | (12,48,1536) | Small |
| Green | (6,192,768) | (30,90,384) | (12,48,1536) | |

come up, it is immaterial how the numbers are distributed around the faces.

Now, rolling the 9 large dice is more likely to yield a greater total than the 9 medium dice; the 9 medium dice are more likely to yield a higher sum than the 9 small ones, but the 9 small ones are more likely to give a greater sum than the 9 big ones.

Besides this, we have that, nested within each size category, the 3 red dice are more likely to produce a greater total than the 3 yellow ones, while the yellow ones are likely to do better than the green ones, and the green dice are likely to beat the red ones.

Finally, nested within each set of 3 of a given size and color, the die marked $A$ is likely to give a greater result than $B$, $B$ greater than $C$, and $C$ is likely to be better than $A$!

Perhaps the really amazing thing is (as Funkenbusch and Saari demonstrate) that such nesting can be carried to any level whatsoever, not just through 3 levels of size, color and label.

As overwhelming as all this might be, Walter is not done yet. He is now ready with a magnificent coup de grâce. Walter has succeeded in going ISOTROPIC!! The following set of 27 dice, distinguished by size, color and label as before, not only enjoys all the intransitivity of the set just described, but contains an intransitive cycle for *every* intersection of 2 or more of the classes of dice. Whenever the distinction between the opposing subsets of dice is one of size, the cyclic order of preference is large over medium over small over large; for colors the preference is red-yellow-green-red, and for letters it is $A$-$B$-$C$-$A$. Thus, for example,

> the 3 large red dice are not only preferable to
> the 3 large yellow ones (red over yellow), but also to
> the 3 medium red ones (large over medium);
> the 3 yellow $A$'s are preferable to the 3 green $A$'s;
> the small green $C$ is to be preferred to the small red $C$
> (and also to the large green $C$ and the small green $A$).

You can vary whatever characteristic you like in order to enter an intransitive cycle of similar subsets. Unfortunately, this cannot be extended to an arbitrary number of classifications; but, if one restricts oneself to just 3 categories, there does exist an isotropic set of $3^n$ dice for each positive integer $n$.

This latest development is so recent (May, 1984) that, at the time of writing, it has not been described elsewhere. However, the interested reader can look for further discussion of this subject in future papers of Walter Funkenbusch and Donald Saari.

## Exercises

1. In section 1, show that the odds are 2:1 that $R$ (red) will win against $W$ (white); 5:4 that $W$ will win over $B$, and 2:1 that $B$ will win against $R$.

2. In section 2, show that the preferences reverse when two copies of the selected dice are rolled. In particular, show that the odds are

## A THIRD LEVEL ISOTROPIC NEST

**Large**

| | *A* | *B* | *C* |
|---|---|---|---|
| Red | (3,42,360) | (2,41,362) | (4,40,361) |
| Yellow | (2,41,362) | (1,40,364) | (3,39,363) |
| Green | (4,40,361) | (3,39,363) | (5,38,362) |

**Medium**

| | *A* | *B* | *C* |
|---|---|---|---|
| Red | (2,41,362) | (1,40,364) | (3,39,363) |
| Yellow | (1,40,364) | (0,39,366) | (2,38,365) |
| Green | (3,39,363) | (2,38,365) | (4,37,364) |

**Small**

| | *A* | *B* | *C* |
|---|---|---|---|
| Red | (4,40,361) | (3,39,363) | (5,38,362) |
| Yellow | (3,39,363) | (2,38,365) | (4,37,364) |
| Green | (5,38,362) | (4,37,364) | (6,36,363) |

4:5 that $2R$ will win against $2W$, 11:16 that $2W$ will win over $2B$, and 4:5 that $2B$ will win against $2R$.

3. In section 3, show that the preferences (established in section 1) are again reversed by adding the appropriate one of $X$, $Y$, $Z$. In particular, show that the odds of $R + X$ winning over $W + X$ are 13:14, those of $W + Y$ beating $B + Y$ are 8:9, and the odds of $B + Z$ winning over $R + Z$ are 13:14.

4. In section 4, show that the odds of the one orange die beating one green die are 2:1, of two $O$'s beating $2G$'s are 5:4, but of three $O$'s beating $3G$'s are 53:55.

Clearly, then, with only 2 of each color to choose from, Walter can always gain the advantage by choosing orange.

5. In section 5, show that the odds that $P$ comes out ahead of the combination of $Q$ and $R$ are 14:13, that $Q$ beats the combination of $P$ and $R$ are 67:41, and that $R$ beats $P$ and $Q$ are 37:17.

### Reference

1. W. W. Funkenbusch and D. G. Saari, Preferences among preferences, or, nested cyclic stochastic inequalities, Congressus Numerantium, 39 (1983) 419–432.

CHAPTER 5

# TWO PROBLEMS IN GRAPH THEORY

## 1. A Problem on Weighted Trees

In graph theory, a tree is a graph that is all in one piece and has no cycles. One of the most basic properties of a tree is that a path along its edges between two vertices is always unique. By definition, a path cannot repeat a vertex, thus prohibiting the retracing of an edge, and if one were to encounter a choice of paths at some vertex en route from vertex $u$ to vertex $v$, a cycle would be inevitable for the paths would have to come together again by the time they reach the final destination, if not sooner. Since there is exactly one path between each pair of vertices, a tree with $n$ vertices defines a total of $\binom{n}{2}$ different paths.

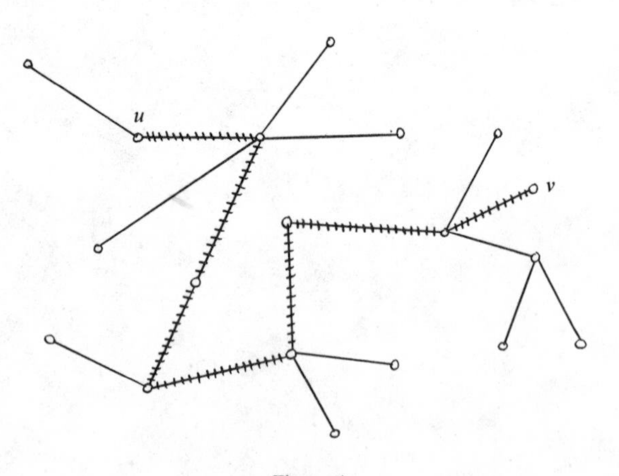


Figure 1

Now suppose that each edge of a tree is weighted with some positive integer and that we take the weight of a path to be the sum of the weights of its edges. For example, consider the weighted trees in Figures 2 and 3. In Figure 2, the number of vertices is 4, the number of paths is $\binom{4}{2}=6$, and their weights are

$$AB = 1, \quad BC = 2, \quad AC \text{ (i.e., } ABC) = 1 + 2 = 3,$$
$$BD = 4, \quad AD = 5, \quad \text{and} \quad CD = 6.$$

In Figure 3, the tree has 6 vertices and $\binom{6}{2} = 15$ paths, whose weights are 1, 2, 3, 4, ..., 15. We can't help noticing that in both these cases the $\binom{n}{2}$ paths have as weights the first $\binom{n}{2}$ positive integers 1, 2, ... $\binom{n}{2}$.

It is natural to wonder whether there exists such a weighted tree for every value of $n$.

> Does there exist, for each positive integer $n > 1$, a weighted tree having $n$ vertices with paths of weights 1, 2, ..., $\binom{n}{2}$?

The cases of $n = 2$ and 3 are trivial (Figure 4). It might come as a mild surprise, however, to learn that, besides the above solutions for $n = 4$ and 6, no other such weighted tree has ever been found. In general, there are many different trees having $n$ vertices, and to solve our problem we need to come up with the happy combination of a feasible shape of tree and a weighting of the edges that suits it.

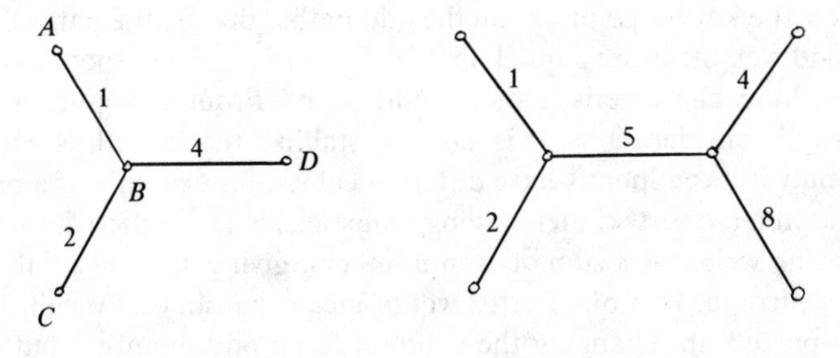


Figure 2 Figure 3

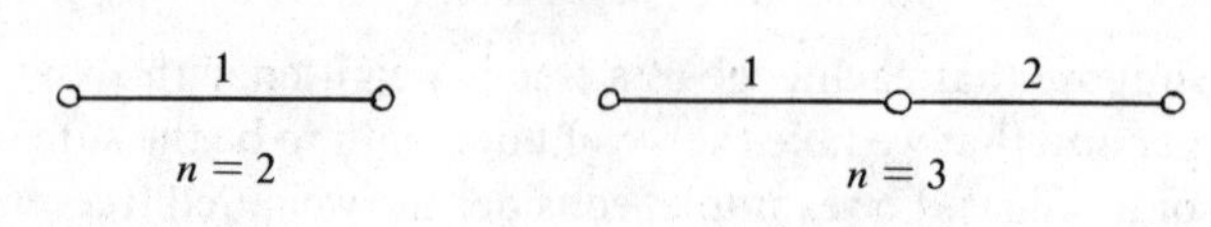


Figure 4

It would be too bad to waste a lot of time on a tree that was doomed from the start. Thus it was of particular interest in 1977 when Herbert Taylor (University of Southern California) proved that there is no use trying any tree having 5 vertices, or 7, or 8, 10, 12, 13, 14, 15, 17, 19, 20, 21, 22, ....

TAYLOR'S CONDITION. *A value of $n$ cannot possibly work unless it is a perfect square or a perfect square plus* 2.

His easy, beautiful proof is truly a gem [**1**].

Let $T$ be any successfully weighted tree; suppose it has $n$ vertices. Now, walk through $T$, coloring the vertices either red or blue, as follows. Start at any vertex with either color and proceed through $T$ to the other vertices, switching colors only after traversing an edge that has an *odd* weight (thus every edge having an even weight will have both ends the same color and every edge having an odd weight will have ends of different colors). Suppose $R$ of the vertices get colored red, and $B$ get colored blue. Since every vertex is either one color or the other, we have

$$R + B = n.$$

Now the key to the proof are the odd paths, that is, the paths having odd weight. From a quick look at Figure 5, can you spot the distinguishing characteristic of an odd path? From straightforward "parity" considerations, it is easy to establish that a path is odd if and only if its endpoints have different colors: for example, if a path begins at a red vertex, then so long as it sticks to red vertices its accumulating weight is a sum of even numbers, giving an even subtotal; the occurrence of a blue vertex would mean that an odd weight has been picked up, changing the subtotal to an odd quantity; but, so long as the path remains on blue vertices, additional even-weighted

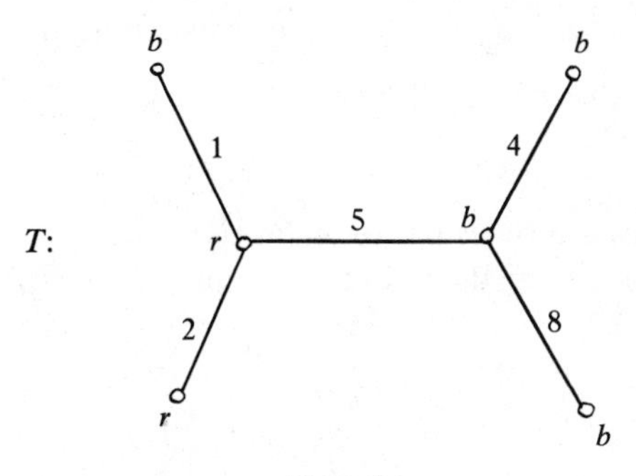


Figure 5

edges keep the growing total at an odd amount; clearly the subtotal seesaws between even and odd values at each change of color.

Since each of the $R$ red vertices can be coupled with each of the $B$ blue ones to give a pair of unlike endpoints, we have that

$$\text{the number of odd paths} = RB.$$

But the weights of the paths in $T$ are the numbers $1, 2, \ldots \binom{n}{2}$, generally half of which are odd.

(a) *If* $\binom{n}{2}$ *is even*, then exactly half the paths are odd, and we have

$$RB = \frac{1}{2}\binom{n}{2} = \frac{n(n-1)}{4},$$

$$4RB = n^2 - n,$$

$$n = n^2 - 4RB.$$

Recalling that $R + B = n$, we can rewrite the right side to give

$$n = (R + B)^2 - 4RB = (R - B)^2,$$

a perfect square.

(b) *If* $\binom{n}{2}$ *is odd*, there is one more odd than even path, and we have

$$RB = \frac{1}{2}\left[\binom{n}{2} + 1\right],$$

which, in a similar way, yields $n = (R - B)^2 + 2$.

Further information and references concerning this problem are given in [2].

## 2. A Packing Problem

A graph with $n$ vertices has room for $\binom{n}{2}$ edges, and one which possesses all $\binom{n}{2}$ possible edges is called a *complete* graph, denoted by $K_n$.

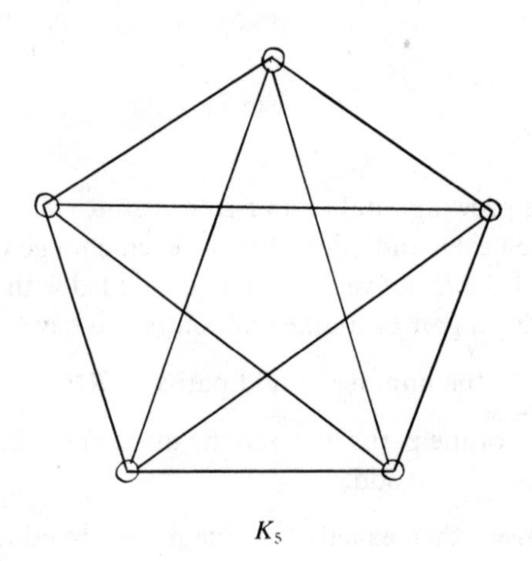

$K_5$

While there is only one way to construct a tree with one edge or two edges, there are two trees with 3 edges and, in general, the family of trees having a specified number of edges has many different members. If one selects a tree from each of the families with 1, 2, 3, ..., $n-1$ edges, a collection of $n-1$ trees is obtained in which the total number of edges is

$$1 + 2 + \cdots + (n-1) = \frac{(n-1)n}{2} = \binom{n}{2}.$$

Since this is the number of edges in a $K_n$, the question arises whether such a set of $n-1$ trees can always be packed into a $K_n$ so as to cover all the edges (naturally, the trees are not to be dismembered and the packing is to be done vertex-on-vertex and edge-on-edge). At present, this is unsolved for the general case in which one is free to taken an arbitrary tree from each family. However, each family has a member which is merely a path and another member which is a star (these coincide for the family of trees with 1 edge and for the family of trees with 2 edges). It is our purpose here to show that if one is restricted to selecting either the path or the star from each family, then there is always a way to accomplish the packing (at each family the choice is independent, allowing any mixture of paths and stars). The following proof is due to S. Zaks and C. L. Liu (University of Illinois at Urbana-Champaign) **[3]**.

Suppose the $n$ vertices of a $K_n$ are numbered 1, 2, ..., $n$. As noted, a $K_n$ contains an edge for each of the $\binom{n}{2}$ pairs of vertices. Now, $\binom{n}{2}$ cells can be arranged in $n - 1$ rows and columns, as shown in Figure 7, with the rows numbered 1 through $n - 1$ and the columns 2 through $n$. Taking the cell in row $i$ and column $j$ to represent the edge joining vertices $i$ and $j$, the packing of a $K_n$ corresponds to the packing of this triangular array. This is a brilliant way of transforming the problem, for the conclusion now follows immediately by induction.

Since a path with one or two edges is identical, respectively, to a star with one or two edges, for $n = 4$ there are only two possible cases, depending on whether the path or the star is chosen from the family of trees having three edges. Figure 8 shows how to pack the triangular array in each of these cases (the vertices are numbered 1 through 4 and the edges of the different trees are distinguished by the letters $a$, $b$, $c$).

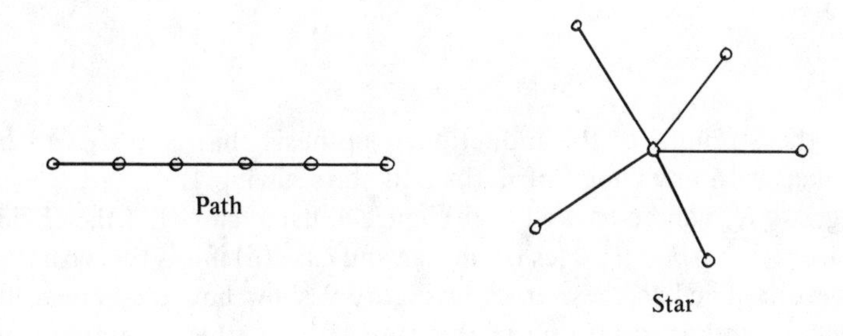


Figure 6

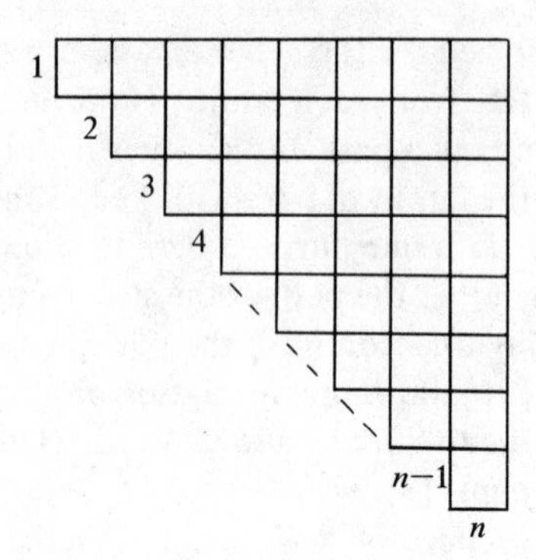


Figure 7

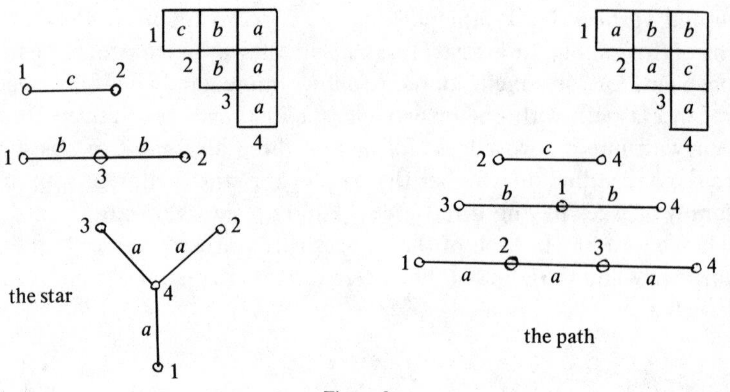


Figure 8

On the strength of the induction hypothesis that a $K_{n-1}$ can be packed with any choice of paths and stars having 1, 2, ..., $n - 2$ edges, a $K_n$ can be packed as in Figure 9, using case (i) if the choice of tree having $n - 1$ edges is the star and case (ii) if it is the path (the labels assigned to these trees in Figure 9 show how they are to be packed; and after putting in this star or path, the remaining cells constitute exactly the triangular array for a $K_{n-1}$, and the remaining trees are appropriately those having 1, 2, ..., $n - 2$ edges;

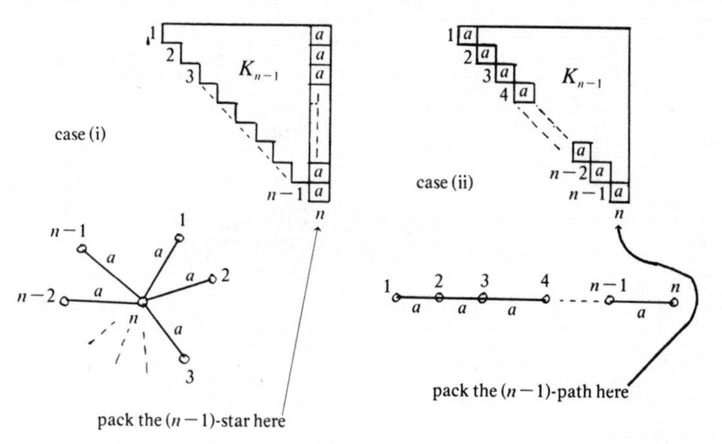


Figure 9

thus the packing can be completed in accordance with the induction hypothesis).

Several interesting similar problems are also solved in [**3**].

## References

1. Herbert Taylor, Odd path sums in an edge-labelled tree, Mathematics Magazine, 50 (1977) 258–259.
2. John Leech, Another tree labelling problem, Amer. Math. Monthly, 82 (1975) 923–925.
3. S. Zaks and C. L. Liu, Decomposition of graphs into trees, Proc. of the Eighth Southeastern Conference on Combinatorics, Graph Theory, and Computing, 1977, pp. 643–654.

CHAPTER 6

# TWO APPLICATIONS OF GENERATING FUNCTIONS

## 1. A Theorem of Euler

A partition of a positive integer $n$ is simply a way of expressing $n$ as an unordered sum of positive integers. The number of partitions of $n$ is denoted by $p(n)$. For example, $p(4) = 5$ because of the 5 expressions

$$4, \qquad 3+1, \qquad 2+2, \qquad 2+1+1, \qquad 1+1+1+1.$$

The function $p(n)$ increases rapidly with $n$. For example, $p(10) = 42$, $p(20) = 627$, $p(50) = 204226$, and $p(200)$ is almost 4 trillion. The only known formula for $p(n)$ is a complicated infinite series which does not readily lend itself to elementary investigations. Happily, we can establish many things without an explicit formula.

Naturally there are many distinguishing features that can be used to classify the various partitions of a given positive integer $n$. For example, we might wish to study the partitions which contain only even numbers, or those whose smallest term is at least a certain minimum. We denote the *number* of partitions of $n$ in the class distinguished by a property $S$ by the symbol $p(S, n)$. Thus, if $O$ denotes the property of containing *only odd* parts, then $p(O, n)$ is the number of ways of writing $n$ as a sum of odd positive integers, repetitions allowed, of course. From the partitions of 4, listed above, we see that

$$p(O, 4) = 2 \qquad (\text{counting} \quad 3+1 \quad \text{and} \quad 1+1+1+1).$$

Another class of interest contains the partitions of $n$ in which all the parts are *different*. Calling this class $D$, we have

$$p(D, 4) = 2 \qquad (\text{for} \quad 4 \quad \text{and} \quad 3 + 1).$$

The 7 partitions of $n = 5$ are

$$5, \quad 4 + 1, \quad 3 + 2, \quad 3 + 1 + 1, \quad 2 + 2 + 1,$$
$$2 + 1 + 1 + 1, \quad 1 + 1 + 1 + 1 + 1.$$

Thus we have

$$p(O, 5) = 3 \qquad (\text{for} \quad 5, \quad 3 + 1 + 1, \quad \text{and} \quad 1 + 1 + 1 + 1 + 1),$$

$$p(D, 5) = 3 \qquad (\text{for} \quad 5, \quad 4 + 1, \quad \text{and} \quad 3 + 2).$$

It is curious that $p(O, n)$ and $p(D, n)$ are equal for $n = 4$ and 5. For $n = 6$, we find

$$p(O, 6) = 4 \qquad (\text{for} \quad 5 + 1, \quad 3 + 3, \quad 3 + 1 + 1 + 1, \quad 1 + 1 + 1 + 1 + 1 + 1),$$

$$p(D, 6) = 4 \qquad (\text{for} \quad 6, \quad 5 + 1, \quad 4 + 2, \quad 3 + 2 + 1).$$

In 1748 the great Euler made the amazing discovery that these numbers are always the same! It staggers the intuition that such dissimilar conditions, "odd" and "different," should always lead to the same number of partitions. This is far from being the only surprise in this fascinating area.

THEOREM. *$p(O, n) = p(D, n)$ for all positive integers $n$.*

*Proof.* (a) One proof consists of a very simple application of generating functions. It is not necessary that the reader have an acquaintance with generating functions; the applications we shall discuss are self-contained.

It is easy to convince oneself that the coefficient of $x^n$ in the following product is just the number $p(O, n)$:

$$f(x) = (1 + x + x^2 + x^3 + \cdots)(1 + x^3 + x^6 + x^9 + \cdots)(1 + x^5 + x^{10} + x^{15} + \cdots)\cdots$$

$$= \prod_{k=1}^{\infty} (1 + x^k + x^{2k} + x^{3k} + \cdots + x^{nk} + \cdots),$$

where $k$ is odd.

For example, in constructing a term in $x^{23}$ by choosing $x^2$ from the first bracket, $x^6$ from the second, $x^{15}$ from the third, and 1 from every other bracket, one builds up the exponent 23 as the sum

$$2 + 6 + 15,$$

which can be interpreted as the partition of 23 given by the odd integers

$$1 + 1 + 3 + 3 + 5 + 5 + 5$$

(the exponent of the term taken from the first bracket indicates the number of 1's in the partition, the exponent from the second bracket the number of 3's, and so on). Conversely, each such partition of 23, used as a prescription for the selection of terms from the brackets in $f(x)$, will yield a term having exponent 23.

Since the infinite series $1 + x^k + x^{2k} + \cdots$ is just $(1 - x^k)^{-1}$, we can write $f(x)$ in the form

$$f(x) = \prod_{k=1}^{\infty} \frac{1}{1 - x^k} = \frac{1}{1-x} \cdot \frac{1}{1-x^3} \cdot \frac{1}{1-x^5} \cdots \frac{1}{1-x^k} \cdots \quad (k \text{ odd}).$$

On the other and, we see similarly that $p(D, n)$ is the coefficient of $x^n$ in the expansion of the product $g(x)$:

$$g(x) = (1 + x)(1 + x^2)(1 + x^3)(1 + x^4) \cdots (1 + x^k) \cdots.$$

In $g(x)$ there is a bracket for each integer 1, 2, 3, ..., but instead of using the entire series $1 + x^k + x^{2k} + \cdots$, as in $f(x)$, we truncate the series at the second term to yield $(1 + x^k)$ in order to offer for this bracket only the two choices of selecting either the 1, which contributes zero to the accumulating exponent of $x$, or the $x^k$, which contributes $k$ to it. Clearly there is no way for a nonzero contribution $k$ to be repeated. Thus each term in $x^n$ displays $n$ as a sum of distinct positive integers. And since each partition of $n$ into distinct parts provides a prescription for producing one term in $x^n$, it follows that $p(D, n)$ is indeed the coefficient of $x^n$ in $g(x)$.

Now, the typical factor $(1 + x^k)$ in $g(x)$ may be rewritten in the form

$$(1 + x^k) = (1 + x^k)\frac{1 - x^k}{1 - x^k} = \frac{1 - x^{2k}}{1 - x^k}.$$

Therefore we have

$$g(x) = \frac{1 - x^2}{1 - x} \cdot \frac{1 - x^4}{1 - x^2} \cdot \frac{1 - x^6}{1 - x^3} \cdot \frac{1 - x^8}{1 - x^4} \cdots.$$

In the denominator we have a factor $1 - x^k$ for *every* $k = 1, 2, 3, \ldots$, while in the numerator these same factors occur only for *even* values of $k$. Cancellation, then, yields

$$g(x) = \frac{1}{1 - x} \cdot \frac{1}{1 - x^3} \cdot \frac{1}{1 - x^5} \cdots \frac{1}{1 - x^k} \cdots, \quad \text{for } k \text{ odd.}$$

Therefore the generating functions $f(x)$ and $g(x)$ are really the same, implying the desired conclusion $p(O, n) = p(D, n)$ for all $n$.

*Proof.* (b) While not appearing to be a very promising approach at first, an explicit 1–1 correspondence between these two classes of partitions is achieved easily as follows (easy things like this are often the most difficult to think of in the first place).

First of all, observe that every position integer can be written in the form $2^r q$, where $q$ is odd. Now two integers of this form can be equal, $2^r q = 2^t s$, only if the exponents are equal, $r = t$ (since the odd numbers $q$ and $s$ contain no additional factors 2), and it follows that the odd factors must also be equal, $q = s$.

Let us demonstrate with a particular example that faithfully represents the general case. We begin with the partition of 71 into the odd parts

$$\begin{aligned}&7 + 7 + 7 + 5 + 5 + 5 + 5 + 5 + 3 + 3 + 3\\ &\quad + 3 + 3 + 3 + 3 + 3 + 1.\end{aligned}$$

First gather like terms and write the coefficients in the binary scale (that is, as the sum of *distinct* powers of 2):

$$\begin{aligned}71 &= 3(7) + 5(5) + 8(3) + 1(1)\\ &= (2^1 + 2^0) \cdot 7 + (2^2 + 2^0) \cdot 5 + (2^3) \cdot 3 + (2^0) \cdot 1.\end{aligned}$$

Now multiply out to get terms of the form $2^r q$, where $q$ is odd, and simplify:

$$\begin{aligned}71 &= 2^1 \cdot 7 + 2^0 \cdot 7 + 2^2 \cdot 5 + 2^0 \cdot 5 + 2^3 \cdot 3 + 2^0 \cdot 1\\ &= 14 + 7 + 20 + 5 + 24 + 1\\ &= 24 + 20 + 14 + 7 + 5 + 1.\end{aligned}$$

No two of these terms could be equal because the powers of 2 that combine with a given odd number $q$ are all different.

Since this procedure is reversible, the desired correspondence is established.

We observe that this theorem of Euler's is just a special case of exercise 2 (given by $d = 1$). There are many surprises in this remarkable subject, as is evidenced by the striking results contained in the exercises below.

The standard work on partitions is an extensive study by George Andrews (Pennsylvania State University)—*The Theory of Partitions, Encyclopedia of Mathematics and Its Applications*, Volume 2 (Addison Wesley), 1976. A recent book—*Combinatorial Enumeration*—by my colleagues David Jackson and Ian Goulden (Academic Press, 1983) contains a wealth of material on generating functions. The reader might also enjoy the very readable introductory exposition "The Use of Generating Functions to Discover and Prove Partition Identities," by Henry Alder (University of California, Davis), appearing in the *Two-Year College Mathematics Journal* (1979, pp. 318–329).

**Exercises**

1. The number of partitions of $n$ in which *no even part is repeated* (the odd parts may or may not repeat) is the same as the number of partitions of $n$ in which *no part occurs more often than* 3 *times*, and is also the number of partitions of $n$ in which *no part is divisible by* 4.

2. The number of partitions of $n$ in which *no part occurs more often than d times* is the same as the number of partitions of $n$ in which *no term is a multiple of* $(d + 1)$.

3. The number of partitions of $n$ in which *each part appears either* 2, 3, or 5 *times* is the same as the number of partitions of $n$ in which *each part is congruent, modulo* 12, *to either* 2, 3, 6, 9, *or* 10.

4. The number of partitions of $n$ in which *no part appears exactly once* is the same as the number of partitions of $n$ in which *no part is congruent to* 1 *or* 5, *modulo* 6.

**2. André's Problem**

There are $5! = 120$ permutations of the numbers 1, 2, 3, 4, 5. Let us scan each of these permutations from left to right, noting as we go from number to number whether we are rising to a bigger number or falling to a smaller one. The sequence of "rises" and "falls" given by the permutation (2 3 5 4 1) is (rise, rise, fall, fall), while the sequence for (2 4 1 5 3) is (rise, fall, rise, fall).

In 1879, the French mathematician D. André solved the challenging problem of determining the number of permutations of $1, 2, \ldots, n$ which provide a sequence of increments that begins with a rise and thereafter alternates fall, rise, fall, rise,... [**1**]. At this point I will not disclose the answer, but will simply alert the reader that one of those "astonishing relations" is in store for him. The 16 such permutations of 1, 2, 3, 4, 5 are

(13254), (14253), (14352), (15243),
(15342), (23154), (24153), (24351),
(25143), (25341), (34152), (34251),
(35142), (35241), (45132), (45231).

Let us try to decompose these special permutations to see if we can discover a suitable way of making them. Since the numbers in these permutations go up and down, they can be represented geometrically by zigzag figures of the following two types:

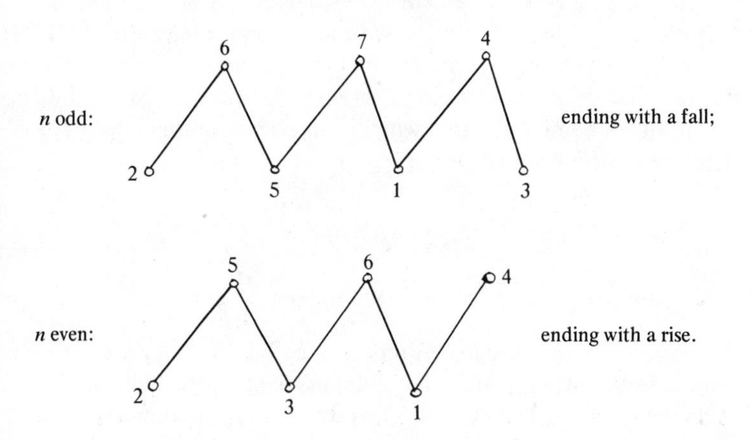

Whether the final increment is a rise or a fall is fundamental to the construction of the permutation. Thus the cases of $n$ odd and $n$ even are sufficiently different to be considered separately. Therefore, for $n$ odd, let $a_n$ denote the number of permutations of the prescribed kind, and for $n$ even, let $b_n$ denote this number.

Now, instead of using an "ordinary" generating function as we did earlier, let us employ the so-called exponential generating function. The difference is simply that the term in $x^r$ carries a factor of $r!$ in its denominator. We need not stop at this point to explain the advantages of doing this; they will become clear as we proceed. Let us simply incorporate the numbers $a_n$ and $b_n$ into the coefficients of the generating functions $f(x)$ and $g(x)$ as follows:

$$n \text{ odd:} \qquad f(x) = \sum_{n=1}^{\infty} a_n \cdot \frac{x^n}{n!} = a_0 + a_1 x + a_2 \frac{x^2}{2!} + \cdots;$$

$$n \text{ even:} \qquad g(x) = \sum_{n=0}^{\infty} b_n \cdot \frac{x^n}{n!} = b_0 + b_1 x + b_2 \frac{x^2}{2!} + \cdots.$$

We have deliberately not omitted the even-powered terms from $f(x)$ and the odd-powered ones from $g(x)$. However, no harm can result, for the coefficients $a_{2k}$ and $b_{2k+1}$, being zero, automatically disappear from any calculation. Since the other initial coefficients represent extreme cases which are not actually realizable, sooner or later we will need to prescribe values for them. We shall see that $a_1 = b_0 = 1$ is not only very convenient but is demanded by our analysis.

Before launching into things, we can observe, in general, that the size of an increment is of no consequence; only its kind, whether a rise or a fall, is of importance. Therefore, it is not the magnitude of the numbers being permuted which counts, but simply how many of them there are; from the mapping

$$\begin{array}{ccccc} 1 & 2 & 3 & 4 & 5 \\ \downarrow & \downarrow & \downarrow & \downarrow & \downarrow \\ 2 & 5 & 11 & 19 & 46 \end{array}$$

it is easy to see that the five numbers 2, 5, 11, 19, 46 give rise to the same number of sequences of increments which (rise, fall, rise, ...) as do the numbers 1, 2, 3, 4, 5. Thus the permutations on every set of

$r$ different numbers contain $a_r$ or $b_r$ of the kind under investigation, as $r$ is odd or even.

In seeking a suitable method of construction, it is generally advantageous to employ building blocks which are the same kind of object as the finished product. This circumstance is obtained if we simply remove from the permutations the maximum element $n$. For the present, let us restrict ourselves to the case of $n$ odd.

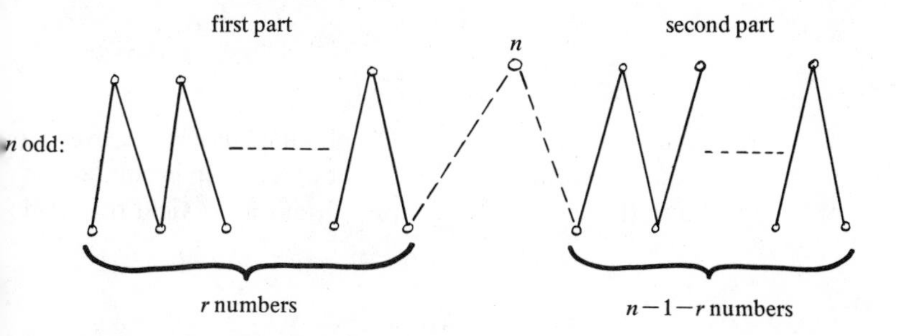


(a) In this case, the sequence will end with a fall. The biggest number must occur in the permutation at the end of a rise and the beginning of a fall, that is, at a "peak" in the corresponding zigzag figure. Let us suppose that its deletion breaks the remaining $n - 1$ numbers into sets of $r$ and $n - 1 - r$ numbers. Because $n$ is odd, the subpermutations in each of these parts must also begin with a rise and end with a fall, implying that each is simply a shorter permutation of odd length. Conversely, each of the $a_r$ possible acceptable permutations of the $r$ numbers in the first part can be connected, via the introduction of the new number $n$, to each of the $a_{n-1-r}$ permutations of the $n - 1 - r$ numbers in the second part to construct a total of $a_r a_{n-1-r}$ acceptable permutations of length $n$. However, as noted above, it doesn't matter which $r$ numbers the first part might happen to contain; every set of $r$ numbers from $1, 2, \ldots, n - 1$ provides $a_r$ acceptable permutations of length $r$ to act as the first part in the construction just described. Since there are $\binom{n-1}{r}$ ways of choosing $r$ integers for the first part, there are a total of

$$\binom{n-1}{r} a_r a_{n-1-r}$$

acceptable permutations of length $n$ in which the number $n$ occurs in the $(r + 1)$th position. We observe, however, that this is precisely the coefficient of the term $x^r x^{n-1-r}/(n - 1)!$ in the square of $f(x)$:

$$[f(x)]^2 = \left(\cdots + a_r \frac{x^r}{r!} + \cdots\right)\left(\cdots + a_{n-1-r} \frac{x^{n-1-r}}{(n-1-r)!} + \cdots\right)$$

$$= \cdots + a_r a_{n-1-r} \frac{(n-1)!}{r!(n-1-r)!} \frac{x^r x^{n-1-r}}{(n-1)!} + \cdots$$

$$= \cdots + \binom{n-1}{r} a_r\, a_{n-1-r} \frac{x^{n-1}}{(n-1)!} + \cdots.$$

The many such terms in $x^{n-1}/(n - 1)!$, obtained as $r$ varies over all possible values, represent the entire set of acceptable permutations of length $n$, and therefore, after simplification, must yield the term

$$a_n \cdot \frac{x^{n-1}}{(n-1)!}.$$

That is to say, pending an investigation of the initial terms in order to ascertain the range of $n$, we have

$$[f(x)]^2 = \Sigma\, a_n \frac{x^{n-1}}{(n-1)!}.$$

We note here that, when $n$ occurs in the second (or second-to-last) position in the permutation, the subpermutations have lengths 1 and $n - 2$, leading to the product $\binom{n-1}{1} a_1 a_{n-2}$, containing the factor $a_1$. In order to account for these cases, then, we are obliged to define $a_1 = 1$.

Now, since $a_0 = 0$ and $a_1 = 1$, the series for $f(x)$ begins $a_1 x + \cdots$, implying that the first term in $[f(x)]^2$ which doesn't vanish is of the second degree. Thus

$$[f(x)]^2 = \sum_{n=3}^{\infty} a_n \frac{x^{n-1}}{(n-1)!} = a_3 \frac{x^2}{2!} + a_4 \frac{x^3}{3!} + \cdots.$$

Since $a_1 = 1$ and $a_2 = 0$, we have $a_1 + a_2 x = 1$, giving

$$1 + [f(x)]^2 = a_1 + a_2 x + a_3 \frac{x^2}{2!} + a_4 \frac{x^3}{3!} + \cdots,$$

which happens to be $f'(x)$, the derivative of $f(x)$.

Taking antiderivatives, then, we obtain the remarkable conclusion that

$$f(x) = \tan x$$

(in $f(x) = \tan x + C$, the value $x = 0$ shows $C = 0$). Therefore, $a_n$ is the coefficient of $x^n/n!$ in the power series for $\tan x$.

(b) Treating the case of $n$ even in a similar fashion, we obtain a second remarkable result. If $n$ is even, the permutation must end with a rise. Since the maximum element $n$ must always occur at the end of a rise, its deletion yields a first part which ends with a fall and a second part that ends with a rise. Therefore, the first part must be a permutation of odd length, while the second part has even length. Consequently, the number of acceptable permutations of length $n$ which have the number $n$ in position $(r + 1)$ is

$$\binom{n-1}{r} a_r b_{n-1-r},$$

containing the factor $b_{n-1-r}$ for the second part. (We observe that, when the number $n$ occurs in the last position in the permutation, the second part has length zero, leading to the product $\binom{n-1}{n-1} a_{n-1} b_0$. Hence the definition $b_0 = 1$ is necessary in order to have these cases included in our total.)

Now, in order to generate these terms $a_r b_{n-1-r}$, the appropriate product of generating functions is

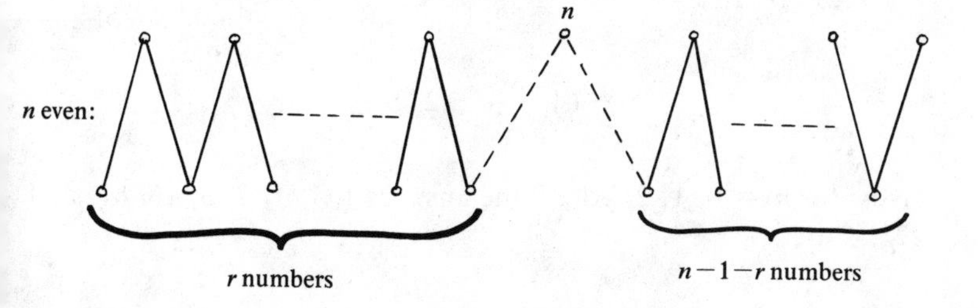

$$f(x)\cdot g(x) = \left(\cdots + a_r\frac{x^r}{r!} + \cdots\right)$$

$$\cdot\left(\cdots + b_{n-1-r}\frac{x^{n-1-r}}{(n-1-r)!} + \cdots\right)$$

$$= \cdots + \binom{n-1}{r} a_r b_{n-1-r}\frac{x^{n-1}}{(n-1)!} + \cdots,$$

simplifying to

$$f(x)\cdot g(x) = \sum_{n=2}^{\infty} b_n \frac{x^{n-1}}{(n-1)!} = g'(x).$$

Since $f(x)$ is known to be $\tan x$, we have

$$\frac{g'(x)}{g(x)} = \tan x,$$

$$\log\,[g(x)] = \int \tan x\, dx = \log(\sec x) + C,$$

where $C$ is a constant. For $x = 0$ we get $C = 0$, and we obtain $g(x) = \sec x$.

Since the power series for $\tan x$ contains only odd-powered terms and the series for $\sec x$ has only even-powered terms, we can roll the two cases together to obtain the single result that, for all $n$, the required number of permutations is the coefficient of $x^n/n!$ in the series for $\sec x + \tan x$. While this is certainly a most satisfying theoretical solution, it must be acknowledged that we have not considered, and will not consider, the problem of calculating the coefficients in this series (they are given in [2]). Let us close with the observation that the first few terms are

$$\sec x + \tan x = 1 + x + \frac{x^2}{2!} + 2\cdot\frac{x^3}{3!} + 5\cdot\frac{x^4}{4!} + 16\cdot\frac{x^5}{5!}$$

$$+ 61\cdot\frac{x^6}{6!} + 272\cdot\frac{x^7}{7!} + \cdots,$$

giving for $n = 0, 1, 2, \ldots, 7$ the answers 1, 1, 1, 2, 5, 16, 61, and 272.

## References

1. D. André, Developpments de sec$x$ et tan$x$. C. R. Acad. Sci., Paris, 88 (1879) 965-967.
2. D. André, Memoire sur le permutations alternées, Journal of Mathematics, 7 (1881) 167-184.

CHAPTER 7

# SOME PROBLEMS FROM THE OLYMPIADS

As a subscriber to *Crux Mathematicorum* [**1**], I have become acquainted with many of the national and international olympiads. These contests and their practice sets are a treasure of good problems. The problems discussed here are ones I have found particularly interesting, but no claim is made that they are the best, or the only good ones, to be found in these contests. Similarly, the given solution, which I hope is at least noteworthy, might not be the easiest or most direct way of solving the problem.

## 1. Austria: 1980—#4

If you add up all the unit fractions (i.e., with numerator 1) having as denominators the products of the members of the nonempty subsets of (1, 2, 3), namely,

$$\frac{1}{1}+\frac{1}{2}+\frac{1}{3}+\frac{1}{1\cdot 2}+\frac{1}{2\cdot 3}+\frac{1}{3\cdot 1}+\frac{1}{1\cdot 2\cdot 3},$$

the total you get is 3. It might come as a pleasant surprise that the fractions generated by $(1, 2, \ldots, n)$ always add up to $n$:

$$\Sigma\frac{1}{a\cdot b\cdots k}=n,$$

*where the sum is taken over all nonempty subsets* $(a, b, \ldots, k)$ of $(1, 2, \ldots, n)$.

The problem is especially easy by induction, which we note in passing. The initial case $n = 1$ is trivial, and if the fractions determined by $(1, 2, \ldots, n-1)$ add up to $n - 1$, it is only necessary to show that the fractions containing $n$ add up to 1. Including $n$ in the denominator of each fraction given by $(1, 2, \ldots, n-1)$, we get a subtotal of

$(1/n)(n-1)$. With $1/n$, itself, then, the sum is indeed 1, and the argument is complete.

Next let us turn to a beautiful solution by David Singmaster (Polytechnic of the South Bank, London, England). Clearly the sum in question is given directly by the expression

$$\left(1+\frac{1}{1}\right)\left(1+\frac{1}{2}\right)\left(1+\frac{1}{3}\right)\cdots\left(1+\frac{1}{n}\right)-1$$

$$=\frac{2}{1}\cdot\frac{3}{2}\cdot\frac{4}{3}\cdots\frac{(n+1)}{n}-1$$

$$=(n+1)-1$$

$$=n.$$

My main interest here, however, is the use of generating functions. The products of the members in the subsets of $(1, 2, \ldots, n)$ determine the coefficients of the function defined by

$$f(x)=(1+x)(1+2x)(1+3x)\cdots(1+nx).$$

(The empty set is represented by the initial term 1 and the entire set by the final term.) Of course, taking a subset from $(1, 2, \ldots, n)$ leaves behind a complementary subset. Dividing each term of $f(x)$ by $n!$ has the effect of moving each subset to the denominator and replacing it with its complementary subset. Thus

$$\frac{f(x)}{n!}=\Sigma\frac{x^t}{a\cdot b\cdots k},$$

where $(a, b, \ldots, k)$ runs over all complementary subsets, which, of course, is just the collection of all subsets. Since the sum we wish to determine concerns only the nonempty subsets, we need to get rid of the term $x^n/1$, corresponding to the empty set. Thus the desired denominators are contained precisely in

$$\frac{f(x)}{n!}-x^n.$$

Unit numerators are achieved neatly by setting $x=1$, and the required sum is given by

$$\frac{f(1)}{n!} - 1 = \frac{2 \cdot 3 \cdot 4 \cdots (n+1)}{n!} - 1$$
$$= (n+1) - 1$$
$$= n.$$

Notice how much these last few steps resemble Singmaster's masterstroke. Although they are admittedly overkill in the present case, generating functions often suggest most elegant ways of looking at things.

## 2. Canada

(a) The International Olympiads are composed from slates of problems contributed by many of the participating countries. Naturally, since there are more problems than can be used, many must be rejected. Here is a Canadian proposal that didn't make it into the 1981 contest.

> The 7 numbers $a$, $b$, $c$, $d$, $e$, $f$, $g$ are nonnegative real numbers that add up to 1. If $M$ is the maximum of the five sums $a+b+c$, $b+c+d$, $c+d+e$, $d+e+f$, $e+f+g$, determine the minimum possible value that $M$ can take as $a$, $b$, $c$, $d$, $e$, $f$, $g$ vary.

One might ponder the situation for quite some time before realizing that this is a job for the pigeonhole principle.* Because the three nonnegative quantities $(a+b+c)$, $(d+e+f)$, $(g)$ add up to 1, they can't all be less than 1/3. With the sum $e+f+g \geq g$, then, $M$ must be at least 1/3. Since the 7-tuple $(a, b, c, d, e, f, g) = (1/3, 0, 0, 1/3, 0, 0, 1/3)$ makes $M$ exactly 1/3, the required minimum is 1/3.

While this certainly settles the matter with dispatch, the following solution, by Jock McKay (University of Waterloo), displays such an elegant application of the pigeonhole principle that it might be preferred to the above.

In the sums that determine $M$, the numbers $a$ and $g$ occur only once, $b$ and $f$ occur twice, and $c$, $d$, and $e$ three times each. If we throw in the numbers $a$, $a+b$, $f+g$, and $g$, then each of $a$, $b$, $c$, $d$, $e$, $f$, $g$ will occur exactly 3 times. Since each of these four extra expressions is just a part of a sum that is already in the expression for $M$, they do not alter the maximum value of the collection of sums. Thus we have

$$M = \max(a+b+c, b+c+d, c+d+e, d+e+f, e+f+g)$$

$$= \max(a, a+b, a+b+c, b+c+d, c+d+e, d+e+f, e+f+g, f+g, g)$$

(containing 9 quantities whose sum is $3(a+b+c+d+e+f+g) = 3$)

$$\geq \text{the average value of } \frac{3}{9} = \frac{1}{3},$$

with equality throughout if and only if all 9 quantities are equal to the average (otherwise the maximum exceeds the average). Thus the required minimum is 1/3, occurring when

$$a = a+b = a+b+c = b+c+d = \cdots = g = \frac{1}{3},$$

which immediately yields the 7-tuple

$$(a, b, c, d, e, f, g) = \left(\frac{1}{3}, 0, 0, \frac{1}{3}, 0, 0, \frac{1}{3}\right).$$

(b) *Canadian Olympiad*: 1982—#3. Determine the smallest number $g(n)$ of points in a set $S$ in $R^n$ such that every point in the space is an irrational distance from at least one point of the set $S$.

It is easy to show that, for the line $R$, the answer is $g(1) = 2$, and this case is left to the reader. It might come as a surprise, however, to learn that, for all greater values of $n$, the answer is the constant $g(n) = 3$.

Settling the case of the plane essentially disposes of all other cases. Consider any pair of points $A$, $B$ on the $x$-axis which are equally spaced on opposite sides of the origin $O$. (Actually any 3 equally spaced points will do.) For an arbitrary point $P(x, y)$, the distances $AP$, $OP$ and $BP$ are connected by the equation

$$AP^2 + BP^2 = 2 \cdot OP^2 + 2 \cdot AO^2.$$

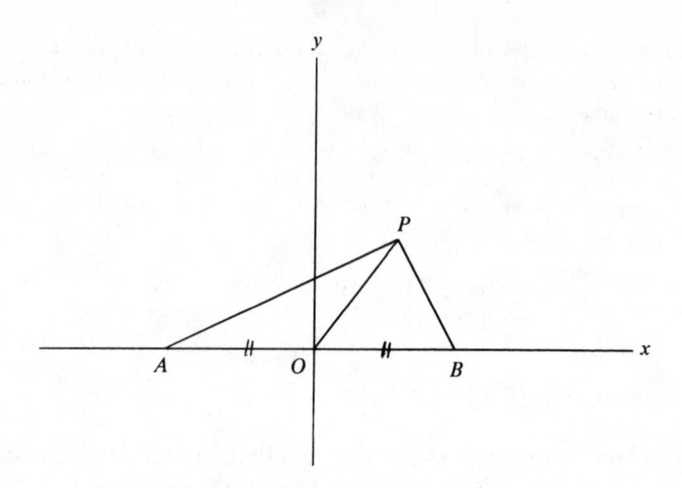


This can be obtained by applying the law of cosines to each of the triangles $AOP$ and $BOP$. However, it is the direct statement of one of the many theorems of the ancient Greek geometer Apollonius [**2**]. It is now evident that a judicious choice of the distance $AO$ will make at least one of the other three distances irrational. Any points which make $AO^2$ irrational will suffice, e.g., $A(-2^{1/4}, 0)$. It follows that $g(2) \leq 3$.

It is left to the reader to convince himself that $g(2) < 3$ is not feasible, yielding the result $g(2) = 3$.

Now every $R^n$ contains three points

$$A(-2^{1/4}, 0, 0, \ldots), \qquad O(0, 0, 0, \ldots), \qquad B(2^{1/4}, 0, 0, \ldots),$$

where all unspecified coordinates are 0's. Since they lie on a straight line (an axis), any fourth point $P$ of the space will determine a plane $PAOB$ in which at least one of the distances $PA$, $PO$, $PB$ is irrational. Thus

$$g(n) = 3 \quad \text{for all } n > 1.$$

### 3. All-Russia

I do not know how many Russian olympiads there are, but the Moscow contests are to be distinguished from the All-Russia competitions. In the All-Russia olympiads, separate contests are held for various grade levels, and I would like to turn now to two questions from the grade 8 and grade 10 papers of 1979–80, questions which concern some very engaging mathematical properties.

(a) *Grade 8*. $ABC$ is a triangle inscribed in a given circle. From a variable point $P$ on the circumference, perpendiculars $PM$ and $PN$ are dropped to $AB$ and $AC$, respectively ($M$ and $N$ occur outside the circle for some positions of $P$). Determine the position of $P$ which makes the length of $MN$ the greatest and find this maximum length.

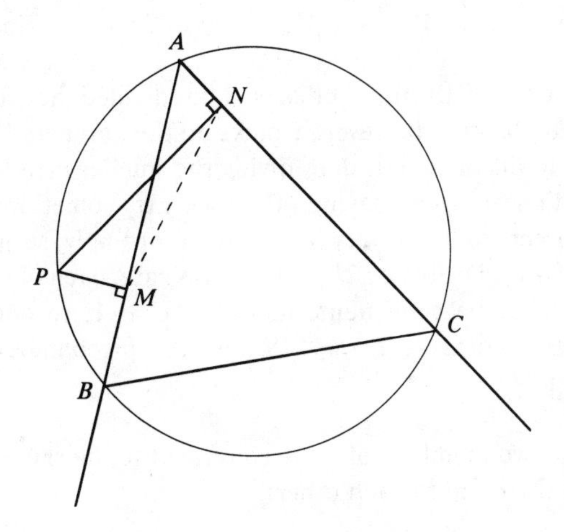


Because of the right angles at $M$ and $N$, a circle $Z$, having $AP$ as diameter, passes through $M$ and $N$. Since $M$ and $N$ are always on $AB$ and $AC$, the chord $MN$ in this circle $Z$ always subtends the same angle $BAC$ at the circumference. As $P$ varies, then, $Z$ changes, but the angle $MAN$ in $Z$ remains the same. Consequently, the greatest chord $MN$ will occur when the circle $Z$ is largest. Clearly this happens when its diameter $AP$ is greatest, that is, when $P$ is diametrically opposite $A$ in the given circle. And, for $P$ in this position, the feet $M$ and $N$ of the perpendiculars will coincide with $B$ and $C$, making the maximum $MN$ equal to the third side $BC$ of triangle $ABC$.

(b) *Grade 10*. For all positive integers $k$, prove that at least one of the integers in the set $S = (2^1 - 1, 2^2 - 1, 2^3 - 1, \ldots, 2^{2k} - 1)$ is divisible by $2k + 1$.

Probably one's first impulse is to check a few easy cases for initial confirmation and the possibility of spotting a useful pattern in the accumulating evidence. Thus we might proceed to make the following table:

| $2k$ | $S$ | $2k + 1$ | Division |
|---|---|---|---|
| 2 | 1, 3 | 3 | $3 \mid 3$ |
| 4 | 1, 3, 7, 15 | 5 | $5 \mid 15$ |
| 6 | 1, 3, 7, 15, 31, 63 | 7 | $7 \mid 63$ |

In each case so far, the number $2k + 1$ has divided the greatest member of $S$. For $2k = 8$, however, $S$ picks up the numbers 127 and 255, but 9 fails to divide 255 (it does divide the smaller member 63 of $S$). This doesn't seem to be paying off, so let's try something else.

The indirect method is very powerful; accordingly, suppose that no member of $S$ is divisible by $2k + 1$. In this case, each of the $2k$ members of $S$ must be congruent, modulo $2k + 1$, to one of the $2k$ nonzero remainders $1, 2, \ldots, 2k$. By the pigeonhole principle,* then, either

(i) some two members of $S$ are congruent to the same remainder, and therefore to each other:

$$2^r - 1 \equiv 2^s - 1 \pmod{2k + 1}, \qquad \text{where } r > s, \qquad \text{or}$$

(ii) each of $1, 2, \ldots, 2k$ is congruent to a different member of $S$.

In the event of case (i), we have

$$2^r - 2^s \equiv 0 \pmod{2k + 1}$$

$$2^s(2^{r-s} - 1) \equiv 0 \pmod{2k + 1}.$$

Since $2k + 1$ is odd, the factor $2^s$ does not contribute toward the satisfaction of the congruence, and it follows that

$$2^{r-s} - 1 \equiv 0 \pmod{2k + 1}.$$

But $2^{r-s} - 1$ *is* a member of $S$, and hence a member of $S$ would be divisible by $2k + 1$.

If (ii) were to hold, then some member of $S$ would be congruent to $2k$: $2^a - 1 \equiv 2k \pmod{2k + 1}$. In this case, then $2^a \equiv 2k + 1 \equiv 0 \pmod{2k + 1}$, an obvious contradiction. The conclusion follows.

## 4. Moscow: Preparation Set—1973

The following two questions are taken from the practice set of problems for the 1973 Moscow contest, not the olympiad paper.

> A point is chosen on each side of a parallelogram $PQRS$ so that the inscribed quadrilateral $ABCD$ that is obtained has area equal to one-half the parallelogram. Prove that at least one of the diagonals of the quadrilateral must be parallel to a side of the parallelogram.

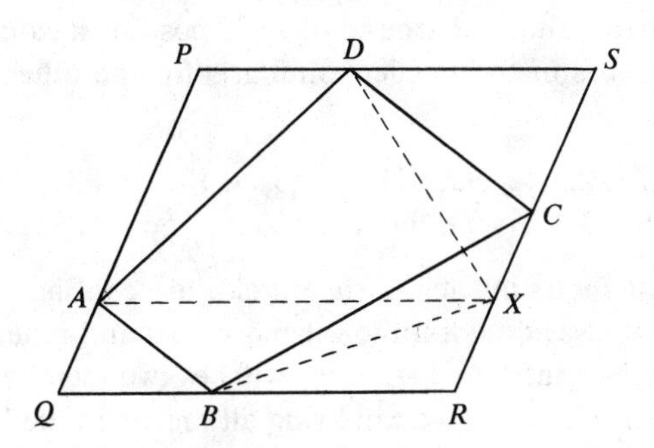


Suppose the diagonal $AC$ were not parallel to the corresponding side $PS$. In this case, construct $AX$ parallel to $PS$ and join $X$ to $B$ and $D$. Then each of triangles $AXD$ and $AXB$ is clearly one-half the parallelogram in which it is inscribed (namely $PAXS$ and $AQRX$), making the inscribed quadrilateral $ABXD$ also one-half the given parallelogram. This quadrilateral shares triangle $ABD$ with $ABCD$, implying that the triangles $BDC$ and $BDX$ have the same area. Accordingly, their common base $BD$ is parallel to the line joining their free vertices $C$ and $X$. Thus, whenever $AC$ is not parallel to $PS$, the diagonal $BD$ is parallel to $RS$.

> The following operation is performed on a 100-digit integer: a block of 10 consecutive integers is chosen at any desired position along the number and the first five digits are interchanged with the last 5 (that is, the segments 1–5 and 6–10 are switched, causing the interchange of digits 1 and 6, 2 and 7, etc.). Two numbers obtained from each other by one or more applications of this operation are called similar. Considering 100-digit integers, composed only of 1's and 2's, what is the maximum number of integers in a collection in which no two are similar?

Numbering the places in an integer 1, 2, . . ., 100 from the left, one sees that the position of any digit that is moved is changed by either $+5$ or $-5$. Thus the 20 digits that occur in positions (1, 6, 11, 16, 21, 26, 31, 36, 41, 46, 51, 56, 61, 66, 71, 76, 81, 86, 91, 96), while capable of being shuffled around these 20 positions, cannot escape to any other places in the integer. Similarly for the other 4 sets of 20 positions:

(2, 7, 12, . . ., 92, 97), (3, 8, 13, . . ., 93, 98),
(4, 9, 14, . . ., 94, 99), (5, 10, 15, . . ., 95, 100).

Although the digits in each of these fields are confined to their own field, these classes do not quite get moved around independently of each other (the shuffling operation switches two members of each of the 5 classes). However, we can bring all the 1's to the leading positions in their classes, thus establishing a standard form for our integers, by operating on them in the following way. (Recall that the digits are only 1's and 2's.)

Let the classes be called 1, 2, 3, 4, 5, according to their initial positions, and let the number of 1's in class $i$ be $n_i$. If $n_i = 0$ for any $i$, then class $i$, consisting only of 2's, is unchanged by all shuffling operations. However, if $n_1 \geq 1$, move a 1 (through whatever intermediate positions may be necessary) to position 1 (the leading position in class 1). Even though class 1 might contain other digits 1, pass on to class 2. If $n_2 \geq 1$, move a 1 to the leading position of class 2 (namely, position 2), and pass on to class 3; if $n_2 = 0$, pass on to class 3 directly. Similarly, move a 1 to the leading position in each of the classes 3, 4, 5, and cycle around through the classes, in order, as often as necessary to bring forward all the 1's in every class. This procedure does not disturb a previously placed 1, and brings the $n_i$ 1's of class $i$ to the first $n_i$ positions in field $i$.

Because the shuffling operation is reversible, it is clear that two integers are similar if and only if they have the same standard form, which simply means that their compositions are described by the same set of "coordinates" $(n_1, n_2, n_3, n_4, n_5)$. Thus the required number of dissimilar integers is the number of different 5-tuples $(n_1, n_2, n_3, n_4, n_5)$, where $0 \le n_i \le 20$, that is, $21^5 = 4084101$.

## 5. West Germany

*1981—#3: First Round*. If a cell is removed from a $2^n \times 2^n$ checkerboard, prove that the remainder can be tiled with $L$-trominos.*

Quartering a $2^n \times 2^n$ board gives 4 squares $2^{n-1} \times 2^{n-1}$. The deleted square must occur in one of these quarters. Placing an $L$-tromino appropriately at the center yields a board in which each quarter has a single square that needs no further attention. Now it is obvious that the conclusion can be obtained easily by induction.

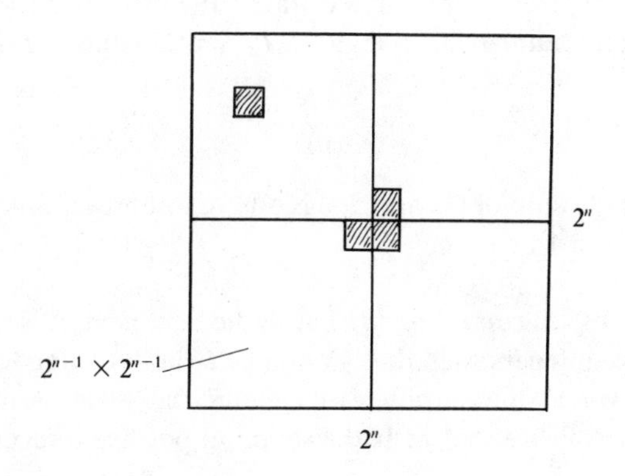

*1982—#3.* $P_1, P_2, \ldots, P_{1982}$ are distinct points in the plane. Prove that every point $P$ of the plane, which does not lie on a segment $P_iP_j$, must occur in the interior of an *even* number of triangles $P_iP_jP_k$.

*Solution* by Jock McKay, University of Waterloo. The trick is to consider the given points in sets of 4, not 3 (for triangles). A set of 4 points determines 4 triangles (perhaps degenerate) and a point $P$, which is not on a side, lies in either 0 or 2 triangles:

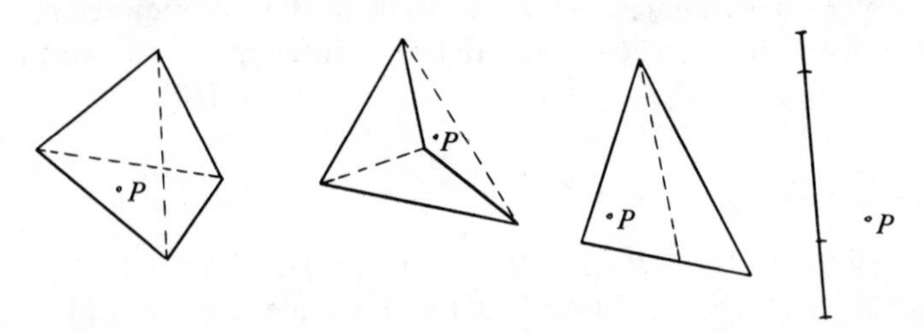


Consider a point $P$ of the plane which is not on a segment $P_iP_j$. This point lies in the interior of 0 or 2 of the triangles determined by the 4 points $(P_1, P_2, P_3, P_4)$, 0 or 2 of the triangles determined by $(P_1, P_2, P_3, P_5)$, and so forth. Adding up the 0's and 2's for all the $\binom{1982}{4}$ possible sets of 4 points $(P_i, P_j, P_k, P_t)$, we obtain a total $N$ that accounts for all triangles that contain $P$, with much duplication, of course. Since a triangle $P_iP_jP_k$ that contains $P$ belongs to the triangles of 1979 quadruples, namely, $(P_i, P_j, P_k, P_t)$ where $P_t$ ranges over the other 1979 given points, it is counted 1979 times in the total $N$. Thus the exact number of triangles $P_iP_jP_k$ that contain $P$ is

$$n = \frac{N}{1979}.$$

Now $N$, being a sum of 0's and 2's, is an even number. Since 1979 is odd, then $n$ must be even.

*1981—#4, Second Round.* Let $M$ be a nonempty set of positive integers such that $4x$ and $[\sqrt{x}]$ both belong to $M$ whenever $x$ does, where $[y]$ denotes the greatest integer $\leq y$. Prove that $M$ is the set of all positive integers.

1. Since $M$ is nonempty, it must contain some positive integer $n$. If $n > 1$, then $\sqrt{n} < n$, and $M$ would also contain the smaller integer $[\sqrt{n}]$. Thus the very smallest integer in $M$ cannot exceed 1, implying that it must, in fact, be 1.

2. Since 1 belongs to $M$, so do 4 and $[\sqrt{4}] = 2$. Because $4x$ belongs to $M$ whenever $x$ does, the presence of 1 brings into $M$ all the even powers of 2: $2^0, 2^2, 2^4, 2^6, \ldots$. Similarly, because 2 belongs to $M$, so do all the odd powers of 2, and we conclude that $M$ contains all powers of 2.

3. Now suppose that some positive integer $n$ fails to belong to $M$. This means that no integer in the half-open interval

$$[n^2, (n+1)^2)$$

could belong to $M$ either, for if some integer $k$ from this interval were to belong to $M$, then so would $[\sqrt{k}] = n$. Similarly, this eliminates from $M$ all the integers $k$ in the range

$$[(n^2)^2, ((n+1)^2)^2)$$

(for such an integer $k$ we would have $n^2 \leq \sqrt{k} < (n+1)^2$, placing $[\sqrt{k}]$ in the range $[n^2, (n+1)^2]$). That is to say, none of the integers in

$$[n^{2^2}, (n+1)^{2^2})$$

belongs to $M$. Continuing this argument, we see that $M$ contains no integer from any range

$$[n^{2^r}, (n+1)^{2^r}), \qquad \text{for } r = 1, 2, \ldots.$$

However, we will see that this is not so, by showing that one of these intervals contains a power of 2 (which we know does belong to $M$).

Clearly

$$\log_2 (n+1) > \log_2 n.$$

Suppose that

$$\log_2 (n+1) = \log_2 n + y.$$

Now, no matter how small the positive quantity $y$ might be, there exists a positive integer $k$ such that $1/2^k < y$. For this integer $k$ we have

$$\log_2 (n + 1) > \log_2 n + \frac{1}{2^k},$$

$$2^k \cdot \log_2 (n + 1) > 2^k \cdot \log_2 n + 1,$$

$$\log_2 (n + 1)^{2^k} > \log_2 n^{2^k} + \log_2 2 = \log_2 2 \cdot n^{2^k}.$$

Taking antilogs, this gives

$$(n + 1)^{2^k} > 2 \cdot n^{2^k}.$$

That is to say, among the intervals which make no contribution to $M$ there is the particular interval $[n^{2^k}, (n + 1)^{2^k})$ which contains completely within it the closed interval $[n^{2^k}, 2 \cdot n^{2^k}]$. But, for all positive integers $m$, the closed interval $[m, 2m]$ contains a power of 2: $m$ itself might be a power of 2; but, if not, then $m$ lies between two powers of 2: $2^r < m < 2^{r+1}$; this yields $2 \cdot 2^r < 2m$, that is, $2^{r+1} < 2m$, placing $2^{r+1}$ in $[m, 2m]$. For this integer $k$, then, the interval does contain an integer in $M$, and the conclusion follows by contradiction.

> *1981—#3, Second Round.* Let $n$ be a power of 2. Prove that, from any set of $2n - 1$ positive integers, one can select a subset of $n$ integers such that their sum is divisible by $n$.

We proceed by induction. The claim holds trivially for $n = 1$, and because every set of 3 integers must contain two which have the same parity (and therefore an even sum), the claim is also valid for $n = 2$.

Suppose that the hypothesis holds for $n = 2^{k-1}$ and that $S$ is a set of $2 \cdot 2^k - 1$ positive integers. First, let the induction hypothesis be applied to any subset of $2 \cdot 2^{k-1} - 1$ members of $S$; this yields a selection of $2^{k-1}$ integers with a sum equal to some multiple of $2^{k-1}$, say $S_1 = 2^{k-1}a$. Removing these integers from the parent collection $S$, we still have a total of $(2 \cdot 2^k - 1) - 2^{k-1} = 3 \cdot 2^{k-1} - 1$ integers left in the remainder of $S$. From any subset of $2 \cdot 2^{k-1} - 1$ of these remaining integers, let a second selection of $2^{k-1}$ numbers which add up to a multiple of $2^{k-1}$, say $S_2 = 2^{k-1}b$, be withdrawn. This still leaves $2 \cdot 2^{k-1} - 1$ numbers in $S$, from which a third such selection of $2^{k-1}$ integers can be made, say with sum $S_3 = 2^{k-1}c$.

Now two of the integers $a$, $b$, $c$ (say, $a$ and $b$) must have the same parity. In this case, the sums $S_1$ and $S_2$, containing a total of $2 \cdot 2^{k-1} = 2^k$ members of $S$ altogether, determine a subset of $2^k$ integers of $S$ that add up to a multiple of $2^k$:

$$\begin{aligned} S_1 + S_2 &= 2^{k-1}a + 2^{k-1}b \\ &= 2^{k-1}(a + b) \\ &= 2^{k-1}(2d) \qquad \text{(for some integer } d\text{, since } a \text{ and } b \text{ have the same parity)} \\ &= 2^k d. \end{aligned}$$

The property established in this problem is, in fact, true for all positive integers $n$, not just for powers of 2. A proof by Ron Graham (Bell Labs) is given in *The Mathematical Intelligencer*, 1979, p. 250.

> *1982—#4.* A set of numbers is called "sum-free" if no two of them, $x$ and $y$ (the same or not), add up to a member $z$ of the set: $x + y = z$. What is the maximum size of a sum-free subset of $(1, 2, 3, \ldots, 2n + 1)$?

This problem is similar to Problem 356, posed by Erwin Just (Bronx Community College), in *Pi Mu Epsilon Journal* (1976, p. 315) and the solution here follows the neat application of the pigeonhole principle* given in the solution published there by Clayton Dodge (University of Maine).

We observe that the segment $(n + 1, n + 2, \ldots, 2n + 1)$ comprises a sum-free subset of $n + 1$ members. We shall show, however, that any subset with $n + 2$ or more members cannot be sum-free.

Let $S = (a, b, \ldots, q)$ denote a subset having $n + 2$ members, the greatest of which is $q$. It is an intriguing result that some two members of $S$ must add up to this maximum element $q$. To see this, pair up the numbers $1, 2, \ldots, q - 1$ in the following way, as far as possible:

$$(1, q - 1), \quad (2, q - 2), \quad (3, q - 3), \quad \ldots.$$

If $q$ is odd, a whole number of pairs is obtained, namely $(q - 1)/2$ of them; if $q$ is even, we get $(q - 2)/2$ pairs with the number $q/2$ left over. In any case, the number of groups produced (counting the singleton in the latter case) is not more than $q/2$. Since

$$q \leq 2n + 1, \qquad \text{we have} \qquad \frac{q}{2} \leq n + \frac{1}{2} < n + 1.$$

In the subset $S$, however, there are $n + 1$ numbers other than $q$, and they are all smaller than $q$. By the pigeonhole principle, then, some two of them must fall into the same group (not the singleton, of course). The conclusion follows.

**6. International Olympiads**

*1981—#5.* Three congruent circles, with centers $X$, $Y$, $Z$ and radius $r$, have a common point $P$ and lie inside a triangle $ABC$. Each circle touches a pair of sides of the triangle. Prove that the incenter $I$ and the circumcenter $O$ of the triangle are in line with the point $P$.

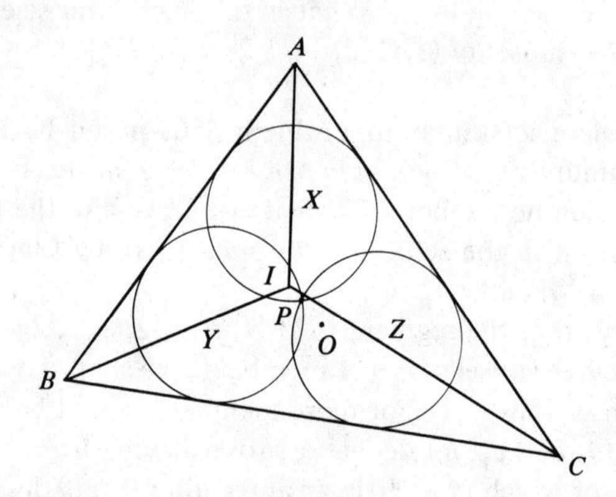


It is well known that $AI$, $BI$, $CI$ bisect the angles of the triangle. But, because of the tangencies, $AX$, $BY$, and $CZ$ also bisect these angles. Thus $IA$ passes through $X$, and $IB$, $IC$ through $Y$ and $Z$.

Because the circles are congruent, $X$ and $Z$ are the same distance from $AC$, making $XZ$ parallel to $AC$. This gives

$$\frac{IX}{IA} = \frac{IZ}{IC}.$$

Similarly, we obtain the three equal ratios

$$\frac{IY}{IB} = \frac{IX}{IA} = \frac{IZ}{IC}.$$

Thus the dilatation* having center $I$ and ratio $IX/IA$ carries $A, B, C$, respectively, to $X, Y, Z$. This means that it would also carry the circle around $ABC$ into the circle around $XYZ$. But dilatations take the center of a circle into the center of the image circle. Therefore the center $O$ is carried to the point $P$, which, being the same distance $r$ from each of $X, Y, Z$, is the center of the circle $XYZ$. It follows that $O$ and $P$ are in line with the center $I$ of the dilatation.

*1979—#1*. Let $p$ and $q$ be positive integers such that

$$\frac{p}{q} = 1 - \frac{1}{2} + \frac{1}{3} - \frac{1}{4} + \frac{1}{5} - + \cdots - \frac{1}{1318} + \frac{1}{1319}.$$

Prove that $p$ is divisible by 1979.

Clearly we have

$$\begin{aligned}\frac{p}{q} &= 1 + \frac{1}{2} + \frac{1}{3} + \cdots + \frac{1}{1319} \\ &\quad - 2\left(\frac{1}{2} + \frac{1}{4} + \frac{1}{6} + \cdots + \frac{1}{1318}\right) \\ &= 1 + \frac{1}{2} + \frac{1}{3} + \cdots + \frac{1}{1319} \\ &\quad - \left(1 + \frac{1}{2} + \frac{1}{3} + \cdots + \frac{1}{659}\right) \\ &= \frac{1}{660} + \frac{1}{661} + \cdots + \frac{1}{1319} \\ &= \left(\frac{1}{660} + \frac{1}{1319}\right) + \left(\frac{1}{661} + \frac{1}{1318}\right) + \cdots \\ &\quad + \left(\frac{1}{989} + \frac{1}{990}\right)\end{aligned}$$

$$= \frac{1979}{660 \cdot 1319} + \frac{1979}{661 \cdot 1318} + \cdots + \frac{1979}{989 \cdot 990}$$

$$= 1979\left(\frac{1}{660 \cdot 1319} + \cdots + \frac{1}{989 \cdot 990}\right).$$

Now 1979 is a prime number. Since each factor in each denominator is less than 1979, the lowest common denominator of these fractions cannot contain a factor equal to 1979. Accordingly, the 1979 in the numerator will survive the simplification to the form $p/q$, implying that 1979 is a factor of the numerator $p$.

*1979—#3*. Two circles $Q$ and $R$ in the plane intersect at $A$ and $Z$. From $A$, a point $S$ goes around $Q$ while a point $T$ traverses $R$. Both points travel in the counterclockwise direction, proceed at constant speeds (not necessarily equal to each other), start together and finish together. Prove the remarkable fact that there exists a fixed point $P$ in the plane with the property that, at every instant of the motions, it is the same distance from $S$ as it is from $T$.

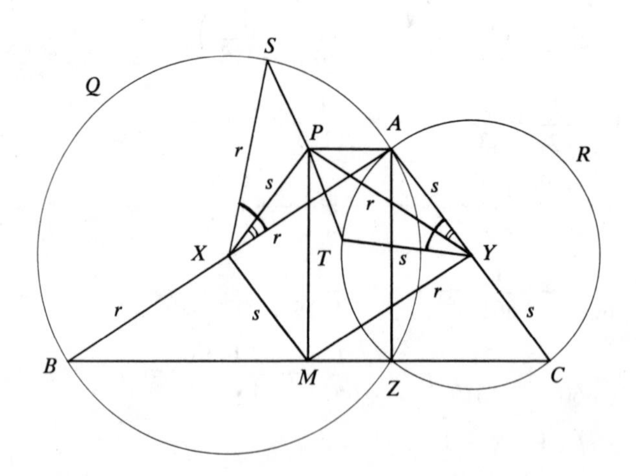

Let the circles $Q$ and $R$ be $X(r)$ and $Y(s)$ (centers $X$ and $Y$, radii $r$ and $s$). If the radii are equal, then clearly $S$ and $T$ are always the same distance from $A$. Suppose, then, that $r > s$.

If $AB$ and $AC$ are diameters, then each of the angles $AZB$ and $AZC$ is a right angle, making $BZC$ a straight segment. Now, when $S$ is at $B$ (halfway around $Q$), $T$ is at $C$, and the proposed point $P$ would have to lie on the perpendicular bisector of $BC$. Let $M$ denote the midpoint of $BC$. By constructing the perpendicular bisector determined by any second pair of simultaneous positions of $S$ and $T$, the possible candidates for $P$ can be narrowed down to the single point of intersection of the two perpendicular bisectors. Even a rough figure strongly suggests that, if such a point exists, it would have to be the point which completes the rectangle having vertices $A$, $Z$, and $M$. This can be proved as follows.

Because $MY$ joins the midpoints of two sides of triangle $ABC$, we have $MY = (1/2)AB = r$. Similarly, $MX = (1/2)AC = s$. Now the line of centers $XY$ is the perpendicular bisector of the common chord $AZ$. Therefore, $XY$ is also the perpendicular bisector of $PM$ (in rectangle $AZMP$). Then, in $XY$ as mirror, $M$ reflects into $P$, and we have $PX = XM = s$ and $PY = YM = r$. As a result, the three sides of triangle $PAX$ are the same lengths as those of triangle $PAY$, making the triangles congruent. This yields

$$\text{angle } PXA = \text{angle } PYA.$$

Let $S$ and $T$ denote an arbitrary pair of simultaneous positions of the moving points. Because the motions proceed at constant speeds and take the same length of time to go all the way around, the points must have the same angular velocity. This makes

$$\text{angle } AXS = \text{angle } AYT.$$

By subtraction, then, we have

$$\text{angle } PXS = \text{angle } PYT,$$

and the triangles $PXS$ and $PYT$ are congruent (two sides and the contained angle). Thus $PS = PT$ and the argument is complete.

**7. Belgium: 1982—#4**

Let a sequence be defined by

$$x_1 = x, \quad \text{and for } k \geq 1, \quad x_{k+1} = x_k^2 + x_k,$$

where $x$ is a real number $\geq 1$. What is the value of the infinite series

$$S = \frac{1}{1 + x_1} + \frac{1}{1 + x_2} + \frac{1}{1 + x_3} + \cdots ?$$

Clearly one is tempted to look at an example or two. Accordingly, for $x = 1$ the sequence begins 1, 2, 6, 42, ..., and the series is

$$S = \frac{1}{2} + \frac{1}{3} + \frac{1}{7} + \frac{1}{43} + \cdots = .99\ldots.$$

If you jump to the conclusion that $S = x_1$, and launch a campaign to prove this, you will be in for a disappointment. For $x = 2$, the sequence and series begin as before, except that the first term of each is missing:

$$S = \frac{1}{3} + \frac{1}{7} + \frac{1}{43} + \cdots.$$

Instead of being near 1, this time $S$ is close to $1/2$, suggesting that $S = 1/x_1$, which conforms to both observed cases. It turns out that this is indeed the case. (The moral is that it is difficult to distinguish between 1 and its reciprocal.)

From

$$x_{k+1} = x_k^2 + x_k = x_k(x_k + 1),$$

we have

$$\frac{1}{1 + x_k} = \frac{x_k}{x_{k+1}}.$$

Thus

$$S = \Sigma \frac{1}{1 + x_k} = \Sigma \frac{x_k}{x_{k+1}}.$$

If we let $S_n$ denote the $n$th partial sum of $S$, we obtain a "telescoping" expression as follows:

$$\begin{aligned}\frac{1}{x_1} - S_n &= \frac{1}{x_1} - \frac{x_1}{x_2} - \frac{x_2}{x_3} - \cdots - \frac{x_n}{x_{n+1}} \\ &= \frac{x_2 - x_1^2}{x_1 x_2} - \frac{x_2}{x_3} - \cdots - \frac{x_n}{x_{n+1}} \\ &= \frac{x_1}{x_1 x_2} - \frac{x_2}{x_3} - \cdots - \frac{x_n}{x_{n+1}} \\ &= \frac{1}{x_2} - \frac{x_2}{x_3} - \cdots - \frac{x_n}{x_{n+1}} \\ &= \frac{1}{x_3} - \frac{x_3}{x_4} - \cdots - \frac{x_n}{x_{n+1}} \\ &\cdots \\ &= \frac{1}{x_n} - \frac{x_n}{x_{n+1}} \\ &= \frac{1}{x_{n+1}}.\end{aligned}$$

It is left to the reader to show that the sequence $\{x_n\}$ is an unbounded sequence of positive numbers (these things are not difficult, but need to be argued carefully), implying that $(1/x_1) - S_n$ approaches zero and the desired $S = 1/x_1$.

## 8. Great Britain: 1982—#6

Prove that the number of 0–1 sequences of length $n$ which contain exactly $m$ occurrences of 01 is $\binom{n+1}{2m+1}$.

I would like to show you two solutions to this beautiful problem. The long expressions in the first one might give the false impression that it is complicated and hard to follow. Actually it is a straightforward application of that great combinatorial tool—generating functions.

(a) Let us use the occurrences of 01 to divide up a sequence into blocks of the following kind: along with a 01 include all consecutive 0's that precede it and all consecutive 1's that follow it, e.g.,

$$\ldots (000\underline{01}11111)(00\underline{01})(0\underline{01}11)(\underline{01})(\underline{01}11) \ldots .$$

In marking off the blocks, one includes *all* preceding 0's until one reaches at 1, not just some of them, and *all* following 1's until one reaches a 0:

$$\ldots 1(\ldots \text{ all preceding 0's } \ldots \underline{01} \ldots \text{ all following 1's } \ldots)0 \ldots .$$

Because of this, the blocks abut along the sequence with no terms between them.

Naturally, the $m$ occurrences of 01 yield $m$ blocks, generally of various lengths. Since these blocks contain the $m$ prescribed occurrences of 01, the only other features that an acceptable sequence could possibly have is an initial segment of consecutive 1's and/or a final segment of consecutive 0's (these are the only things that do not introduce additional occurrences of 01):

$$(\ldots \text{ zero or more 1's } \ldots)\ (\ldots\ m \text{ blocks } \ldots)\ (\ldots \text{ zero or more 0's } \ldots).$$

It might be overstating things a little, but the problem is pretty well solved at this point. All that remains is the introduction of an algebraic representation of this structure and the performance of routine operations to obtain a numerical answer. The actual combinatorics of the problem is already behind us.

Let us represent each 0 by an $x$ and each 1 by a $y$; thus the block 0000011 is represented by $x^5y^2$. We must be prepared to deal with blocks, and consecutive strings, of all lengths. Since $x^i$ denotes the string of $i$ consecutive 0's, the collection of strings of 0's of all lengths is represented by

$$1 + x + x^2 + x^3 + \cdots + x^i + \cdots$$

($x^0 = 1$ denotes the string of length zero). Similarly, the blocks of all lengths are given by the terms $x^iy^j$ in the expansion of $(x + x^2 + \cdots)(y + y^2 + \cdots)$ (at least one 0 and one 1 must be present in a block). Noting that $(1 - x)^{-1} = 1 + x + x^2 + \cdots$, this latter expression is equal to $x(1 - x)^{-1}\,y(1 - y)^{-1}$.

In the same way, a collection of $m$ blocks would be represented by the terms in the expansion of

$$\underbrace{[(x + x^2 + \cdots)(y + y^2 + \cdots)][(x + x^2 + \cdots)(y + y^2 + \cdots)] \cdots [(x + x^2 + \cdots)(y + y^2 + \cdots)]}_{m \text{ times}}.$$

An unsimplified term in this expression has the form

$$(x^a y^b)(x^c y^d) \cdots (x^i y^j),$$

indicating a block containing $a$ 0's and $b$ 1's, followed by a block with $c$ 0's and $d$ 1's, and so forth. The entire expression is given by

$$\begin{aligned}
&[(x + x^2 + \cdots)(y + y^2 + \cdots)]^m \\
&= [x(1 - x)^{-1} y(1 - y)^{-1}]^m \\
&= x^m y^m (1 - x)^{-m}(1 - y)^{-m}.
\end{aligned}$$

To complete the representation, we need only allow for a possible initial segment of consecutive 1's and a trailing segment of consecutive 0's. Thus the acceptable sequences are given precisely by the terms in the expansion of the following generating function $f(x, y)$:

$$\begin{aligned}
f(x, y) = {} & (1 + y + y^2 + \cdots) \\
& [(x + x^2 + \cdots)(y + y^2 + \cdots)]^m (1 + x + x^2 + \cdots).
\end{aligned}$$

This simplifies easily to

$$\begin{aligned}
f(x, y) &= (1 - y)^{-1}[x^m y^m (1 - x)^{-m}(1 - y)^{-m}](1 - x)^{-1} \\
&= x^m y^m (1 - x)^{-m-1}(1 - y)^{-m-1}.
\end{aligned}$$

This function $f(x, y)$ encompasses all acceptable sequences of all lengths. Since each of $x$ and $y$ denotes a term of a sequence (0 or 1), we can focus on the sequences of length $n$ by considering only those terms which contain a total of $n$ $x$'s and $y$'s altogether. Each term of total degree $n$ in $x$ and $y$ represents an acceptable sequence of length $n$, and in its unsimplified form, the term shows exactly how the sequence is constructed. Since we are not concerned with describing the sequences, but just in *counting* them, we don't lose anything by simplifying our expressions. All we really want to know is how many terms of $f(x, y)$ have total degree $n$ in $x$ and $y$ (we don't care whether a term of degree 15 is $x^{11}y^4$ or $x^7y^8$; they each count 1 toward the desired total). In other words, we don't care any more about the difference between $x$ and $y$. In fact, if we changed all the $y$'s to $x$'s, the number we seek would simply become the number of terms $x^n$ in $f(x, y)$ (which would become $f(x, x)$). Accordingly, suppose this is done and that we denote the simplified final coefficient of $x^n$ in a function $g(x)$ by $[x^n]g(x)$. Then the number $N$ of acceptable sequences is given by

$$\begin{aligned} N &= [x^n]f(x, x) \\ &= [x^n]\, x^{2m}\,(1-x)^{-2m-2} \\ &= [x^{n-2m}]\,(1-x)^{-2m-2}. \end{aligned}$$

Now the binomial coefficient $\binom{-n}{r}$ is defined by

$$\binom{-n}{r} = \frac{(-n)(-n-1)\cdots(-n-r+1)}{r!},$$

and it is a routine calculation to show that

$$\binom{-n}{r} = (-1)^r \binom{n+r-1}{r}. \tag{A}$$

Accordingly, we have

$$\begin{aligned} N &= [x^{n-2m}]\,(1-x)^{-2m-2} \\ &= [x^{n-2m}] \sum_{r=0}^{\infty} (-1)^r \binom{-2m-2}{r} x^r \\ &= (-1)^{n-2m} \binom{-2m-2}{n-2m} = (-1)^{2(n-2m)} \binom{n+1}{n-2m} \quad \text{by (A) above} \\ &= \binom{n+1}{n-2m} \\ &= \binom{n+1}{2m+1} \quad \text{since } \binom{k}{r} = \binom{k}{k-r}. \end{aligned}$$

This application might reveal to the reader a new side of generating functions, one in which the emphasis is placed on tailoring a generating function to the problem at hand. This approach requires a certain initial analysis of the constructions involved, but a direct translation of the concrete manipulations often suffices and can lead to a straightforward, transparent solution to a difficult problem.

(b) *Solution* by Ian Goulden, University of Waterloo. As an acceptable sequence $S$ is scanned from left to right, exactly $m$ changes from 0–1 will be encountered. After each of these changes, except the last one, the sequence must switch back to 0's at some point in preparation for the next change from 0–1. Thus, alternating with the $m$ changes from 0–1, there are $m - 1$ changes from 1–0. This collection of $2m - 1$ changes begins and ends with a switch from 0–1. In the event that $S$ begins with a 1, a change from 1–0 must precede the first of the 0–1 switches, and if $S$ were to end in 0, a 1–0 switch would follow the last change from 0–1. Consequently, the number of 1–0 changes might be either $m - 1$, $m$, or $m + 1$. We can guarantee that this number is $m + 1$ simply by putting a 1 in front of $S$ and a 0 after it. This yields an enhanced sequence of length $n + 2$ and guarantees exactly $m$ switches from 0–1 and $m + 1$ switches from 1–0. (If $S$ begins with a 1 on its own, or ends with a 0, these additions add no switches, but in these cases the extra switches are not needed.)

$$1\,|\,\underbrace{11100000111000001101001111110100011}_{\text{S}}\,|\,0$$

An enhanced sequence, then, starts and ends its switches with changes from 1–0 and it has exactly $m + 1$ switches from 1–0 and $m$ switches from 0–1. It follows, then, that an enhanced sequence is completely determined by the disclosure of its $2m + 1$ points of alternation, for each begins with a string of one or more consecutive 1's and thereafter alternates 0's, 1's, 0's, . . . , 0's. Therefore the number of sequences $S$, which is also the number of enhanced sequences (obviously $S$ is recoverable from its enhanced sequence by dropping the leading 1 and the trailing 0), is equal to the number of ways of selecting $2m + 1$ points where the switches are to occur.

Now these switch-points occur *between* consecutive strings in the sequence; they don't actually mark the position of a term in the sequence. In between the $n+2$ positions for the 0's and 1's there are $n+1$ interior spaces that are eligible to be chosen as switch-points. Thus there are $\binom{n+1}{2m+1}$ ways to choose these critical points, and the same number of sequences $S$.

## 9. Hungary: 1981—#1, Third Round

> Six different points are given on a circle. The orthocenter* of the triangle formed by 3 of these points is joined by a segment to the centroid* of the triangle formed by the other 3 points. Prove that the 20 segments that can be determined in this way are all concurrent.

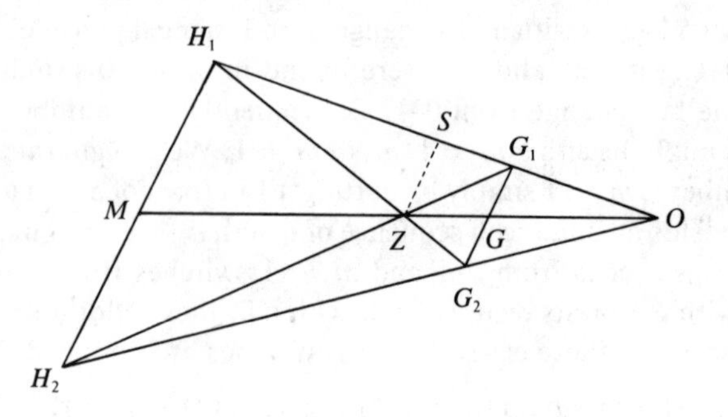


Let $O$ denote the center of the circle and let $H_1$ and $G_1$ be the orthocenter and centroid of one triangle, and $H_2$, $G_2$ the orthocenter and centroid of the complementary triangle. Now for any triangle, the orthocenter $H$, centroid $G$ and circumcenter $O$ lie on a straight line called the Euler line* of the triangle. Since the center $O$ of the given circle is the circumcenter of all the triangles under investigation, each of $H_1G_1$ and $H_2G_2$ is in line with $O$. Not only that, but it is well known (and easy to prove) that $G$ always divides $OH$ in the ratio $1:2$. Thus $G_1G_2$ is parallel to $H_1H_2$.

If $H_1G_2$ and $H_2G_1$ cross at $Z$, then $Z$ must be the required point of concurrency, if indeed it exists. Now, extending $OZ$ to $M$ on $H_1H_2$ completes the quadrangle construction for the midpoint of $H_1H_2$ ([3]) (this result follows from several pairs of similar triangles produced by the parallel segments; the verification of this, and other similar results that follow, are left to the reader). It follows (by similar triangles again) that the point $G$ where $OZ$ crosses $G_1G_2$ is the midpoint of $G_1G_2$. This makes $G$ the center of gravity of a system of equal masses suspended one at each of the 6 given points ($G_1$, $G_2$ are the centers of gravity of the triangles, and $G$ is the center of gravity of $G_1$ and $G_2$). As the chosen triangles change, causing $G_1$ and $G_2$ to vary, the center of gravity of the whole system is not altered. Thus it is the same line $OG$ that is obtained for all choices of triangles.

Again by similar triangles, it is not difficult to establish that the ratio $ZG:GO$ is also $1:2$, showing its independence of the particular choice of triangles (a construction line $ZS$, parallel to $H_1M$ is helpful in showing this). Thus $Z$ is a fixed point and the conclusion follows.

**References**

1. Crux Mathematicorum, Algonquin College, Ottawa, Ontario, Canada.
2. P. A. Petrie, V. E. Baker, W. Darbyshire, J. R. Levitt, and W. B. MacLean, Deductive Geometry and Introduction to Trigonometry, Copp Clark, 1957, pp. 114–115 (Proposition 8).
3. E. A. Maxwell, Geometry for Advanced Pupils, Oxford, 1949, p. 88 (exercise 4).

CHAPTER 8

# A SECOND LOOK AT THE FIBONACCI AND LUCAS NUMBERS

The Fibonacci sequence has been discussed in so many articles and books that it is unlikely you have not encountered the subject elsewhere. One who is not familiar with the topic will find much to enjoy in one of the excellent introductions available, for example [**1**] or [**2**]. Accordingly, the present treatment will touch on the well-known parts of the subject only to gather tools for later use, to add some special comment, or to demonstrate some novel application. The pace will be leisurely, although thorough proofs of many results will be presented. Almost all the ideas here have been gleaned from the splendid journal *The Fibonacci Quarterly*, [**3**], whose steadfast devotion to the subject has shown it to be amazingly fertile.

## 1. The Genealogy of the Male Honeybee

Most people are introduced to the Fibonacci sequence through an entertaining story about breeding rabbits. This is in fact the historical origin of the subject, but it is not nearly as good a story as a more recent one concerning the genealogy of the male honeybee. Ordinary honeybees belong to a class of insects called the hymenoptera, which comprises some 110,000 different species. One of their distinguishing characteristics is a system of controlled reproduction. It seems that early in her career a queen bee goes on a spree, collecting sperm from eager males who, I understand, immediately pass on to their reward (smiling, I trust). Now the queen bee produces many eggs, and it is the general rule that unfertilized eggs hatch into males and fertilized eggs into females. Thus male honeybees do not have fathers. The queen bee is able to store the collected sperm for months and even years and, upon information supplied by her attendants, she can regulate the gender of the offspring to meet the needs of the hive. Female bees are undoubtedly the superior sex; they do everything; in fact, the male's only function is in his role in the production of the prized female. There is no denying that we have here incontestable justification of the warning mothers have tried to pass on to their daughters since time immemorial—men are only good for at most one thing!

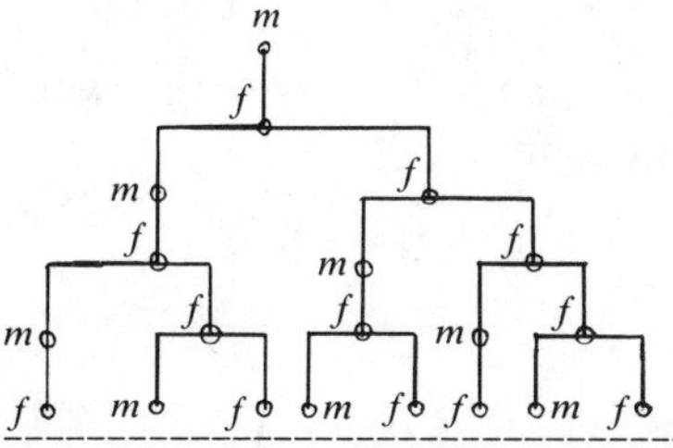

| (m) Male | (f) Female | Total |
|---|---|---|
| (1, | 0) | 1 |
| (0, | 1) | 1 |
| (1, | 1) | 2 |
| (1, | 2) | 3 |
| (2, | 3) | 5 |
| (3, | 5) | 8 |
| (–, | –) | $f_{n-2}$ |
| (–, | $f_{n-2}$) | $f_{n-1}$ |
| ($f_{n-2}$, | $f_{n-1}$) | $f_n$ |

Be that as it may, let us trace the ancestry of a single male honeybee. Let us keep track of things by using $m$ for male and $f$ for female; the totals are given at the right-hand side of the table above. Thus, to begin with, we have our lone male, and the preceding generation consists of a single female. Now his mother had both a mother and a father, and so our bee has two maternal grandparents. Obviously the same parents could have many offspring. However, on the basis that all his ancestors are distinct, we obtain a sequence of totals 1, 1, 2, 3, 5, .... Suppose we let $f_n$ denote the $n$th term in this sequence. Let us say that $f_{n-1}, f_{n-2}, \ldots$ correspond to "previous" generations; although these generations actually exist at a later time, they are encountered earlier in our regressive analysis. Now, everybody has a mother. Therefore, the number of females in a generation is simply the total number of bees in the previous generation. Thus in the $n$th generation there are $f_{n-1}$ females and in the $(n - 1)$th generation there are $f_{n-2}$ females. Since only females have fathers, the number of males in a generation is just the number of females in the previous generation. Thus the $n$th generation has $f_{n-2}$ males to go with its $f_{n-1}$ females, giving a grand total of

$$f_n = f_{n-1} + f_{n-2}.$$

This is the fundamental property of the Fibonacci sequence. In fact, getting away from rabbits and bees, it is customary to define the Fibonacci sequence in terms of this result:

$$f_1 = f_2 = 1, \quad \text{and, for } n > 2, \quad f_n = f_{n-1} + f_{n-2}.$$

It is often convenient to have a value for $f_0$. In keeping with the recursion, we define $f_0$ by the equation

$$f_2 = f_1 + f_0, \quad \text{making} \quad f_0 = 0.$$

Thus the sequence proceeds

$$(0), 1, 1, 2, 3, 5, 8, 13, 21, 34, 55, 89, 144, \ldots.$$

There is no difficulty in extending the sequence backwards indefinitely.

**Exercise**

Prove that $f_{-2n} = -f_{2n}$ and $f_{-(2n+1)} = f_{2n+1}$.

## 2. An Application of the Summation Formula

One of the most well known of all properties of the Fibonacci sequence is the formula for the sum $s_n$ of the first $n$ terms. A glance at the first few cases quickly leads to the conjecture

$$s_n = f_1 + f_2 + \cdots + f_n = f_{n+2} - 1,$$

which is immediately confirmed by mathematical induction. Our present interest in this formula is the following application.

THEOREM. *Let $n$ and $k$ be any two positive integers. Then between the consecutive powers $n^k$ and $n^{k+1}$ there can never occur more than $n$ Fibonacci numbers.*

*Proof.* We proceed indirectly. Suppose the interval between some $n^k$ and $n^{k+1}$ were to contain at least $n + 1$ Fibonacci numbers:

$$n^k < f_{r+1}, f_{r+2}, \ldots, f_{r+n+1}, \ldots < n^{k+1}.$$

Then the sum of the first $n - 1$ of these numbers would be

$$\begin{aligned} f_{r+1} + f_{r+2} + \cdots + f_{r+n-1} \\ &= s_{r+n-1} - s_r \\ &= (f_{r+n+1} - 1) - (f_{r+2} - 1) \\ &= f_{r+n+1} - f_{r+2}. \end{aligned}$$

Solving for $f_{r+n+1}$, we obtain

$$f_{r+n+1} = (f_{r+1} + f_{r+2} + \cdots + f_{r+n-1}) + f_{r+2},$$

which is a sum of $n$ Fibonacci numbers, each of which exceeds $n^k$. Consequently, $f_{r+n+1} > n(n^k) = n^{k+1}$, a contradiction.

## 3. A Geometric Approach

Almost as famous as the formula for $s_n$ is the formula for the sum of the squares of the first $n$ Fibonacci numbers. Again, a glance at the first few cases, followed by mathematical induction, quickly establishes

$$f_1^2 + f_2^2 + \cdots + f_n^2 = f_n f_{n+1}.$$

However, if the method of mathematical induction is a little too sophisticated for someone you would like to tell about this, the following ingenious geometric argument may be used. Consider squares of sides 1, 1, 2, 3, 5, .... Clearly these fit together as indicated in the figure. At each stage of construction, the resulting rectangle has one dimension equal to the side of the last square to be added, say $f_n$, and the other dimension equal to the sum of the sides of the last two squares to be added, namely $f_n$ and $f_{n-1}$. Since $f_n + f_{n-1} = f_{n+1}$, the conclusion follows.

This and many other formulas are deduced geometrically in an article by Alfred Brousseau (*The Fibonacci Quarterly*, 1972, pp. 303-318). Another sample is

$$f_{n+1}^2 = 4f_n f_{n-1} + f_{n-2}^2.$$

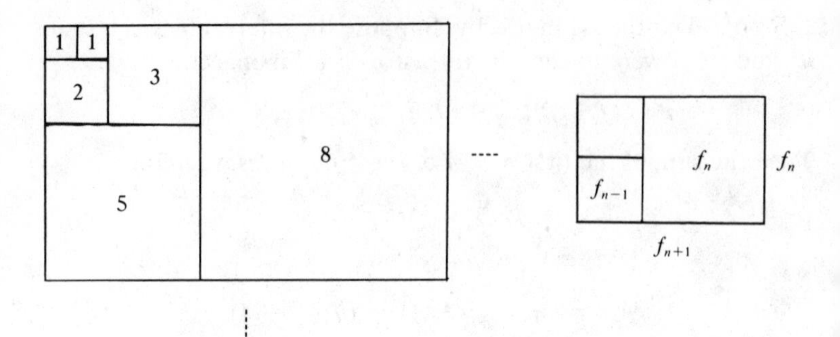


## 4. The Matrix $Q$

A multitude of relations like the ones just described are readily established by mathematical induction. In passing, let us observe how the matrix

$$Q = \begin{pmatrix} 1 & 1 \\ 1 & 0 \end{pmatrix}$$

provides a novel approach to some of these basic results. Actually we have

$$Q = \begin{pmatrix} f_2 & f_1 \\ f_1 & f_0 \end{pmatrix},$$

and it is another easy exercise in mathematical induction to establish the somewhat surprising

$$Q^n = \begin{pmatrix} f_{n+1} & f_n \\ f_n & f_{n-1} \end{pmatrix}.$$

(i) Now, for square matrices $A$ and $B$ of the same size, we have

$$\det AB = (\det A)(\det B), \qquad \text{i.e.,} \qquad |AB| = |A||B|.$$

Thus $|Q^2| = |Q||Q| = |Q|^2$, and, again by induction, it follows that $|Q^n| = |Q|^n$. This translates into

$$f_{n+1}f_{n-1} - f_n^2 = (-1)^n,$$

one of the most useful of Fibonacci identities.

(ii) From the equation $Q^{n+1}Q^n = Q^{2n+1}$, we get

$$\begin{pmatrix} f_{n+2} & f_{n+1} \\ f_{n+1} & f_n \end{pmatrix} \begin{pmatrix} f_{n+1} & f_n \\ f_n & f_{n-1} \end{pmatrix} = \begin{pmatrix} f_{2n+2} & f_{2n+1} \\ f_{2n+1} & f_{2n} \end{pmatrix},$$

which, upon tracing through the multiplication, yields an identity for each Fibonacci number on the right-hand side. For example, for the $f_{2n+1}$ in position 21, we obtain the elegant

$$f_{n+1}^2 + f_n^2 = f_{2n+1}.$$

(iii) Finally, the extremely useful relation

$$f_{n+m} = f_{n-1}f_m + f_nf_{m+1}$$

may be established similarly from the equation $Q^mQ^{n-1} = Q^{n+m-1}$:

$$\begin{pmatrix} f_{m+1} & f_m \\ f_m & f_{m-1} \end{pmatrix} \begin{pmatrix} f_n & f_{n-1} \\ f_{n-1} & f_{n-2} \end{pmatrix} = \begin{pmatrix} f_{n+m} & f_{n+m-1} \\ f_{n+m-1} & f_{n+m-2} \end{pmatrix}.$$

With this formula it is a simple matter to prove by induction the remarkable result that

THEOREM. $f_n | f_{kn}$ *for all* $k = 1, 2, 3, \ldots.$

The claim is obviously valid for $k = 1$. And, using $m = kn$ in the formula, we get

$$f_{(k+1)n} = f_{n+kn} = f_{n-1}f_{kn} + f_nf_{kn+1},$$

which clearly shows that if $f_n | f_{kn}$, then $f_n | f_{(k+1)n}$.

The matrix $Q$ is the subject of Part II of an excellent series of articles in *The Fibonacci Quarterly* entitled "A Primer on the Fibonacci Sequence" (1963, pp. 61-68).

## 5. Binet's Formula

In 1843, the French scholar Binet found a formula giving $f_n$ in terms of $n$. It is a very complicated-looking expression which comes as something of a shock to many of us. The formula is

$$f_n = \frac{1}{\sqrt{5}}\left[\left(\frac{1+\sqrt{5}}{2}\right)^n - \left(\frac{1-\sqrt{5}}{2}\right)^n\right].$$

However, a short, elegant derivation of it can be based on the following lemma, which is immediately established by induction:

LEMMA. *If* $x^2 = x + 1$, *then, for* $n = 2, 3, \ldots,$ *we have*

$$x^n = f_n x + f_{n-1}.$$

The claim is trivial for $n = 2$. And if $x^n = f_n x + f_{n-1}$ for some $n \geq 2$, then

$$\begin{aligned} x^{n+1} &= x^n \cdot x = (f_n x + f_{n-1})x \\ &= f_n x^2 + f_{n-1}x \\ &= f_n(x+1) + f_{n-1}x \\ &= (f_n + f_{n-1})x + f_n \\ &= f_{n+1}x + f_n, \end{aligned}$$

as desired.

Now the two numbers $x$ which make $x^2 = x + 1$ are the roots of the equation $x^2 - x - 1 = 0$, namely

$$\alpha = \frac{1+\sqrt{5}}{2} \quad \text{and} \quad \beta = \frac{1-\sqrt{5}}{2}.$$

Thus, for all $n = 2, 3, \ldots,$ we have

$$\alpha^n = f_n\alpha + f_{n-1}$$

and

$$\beta^n = f_n\beta + f_{n-1}.$$

Subtracting gives $\alpha^n - \beta^n = f_n(\alpha - \beta)$, and, noting that $\alpha - \beta = \sqrt{5}$, we have Binet's formula

$$f_n = \frac{\alpha^n - \beta^n}{\alpha - \beta}.$$

Incidently, to call this "Binet's formula" hardly does justice to Euler and Daniel Bernoulli, each of whom was in possession of it a century earlier. However, Binet did make the discovery independently, and he has so little claim to mathematical fame that it won't hurt these giants to let Binet take the credit for this crumb. It is really surprising that such a seemingly awkward formula can lead to many really easy proofs of quite difficult results. Let us note in passing the simplifying relations

$$\alpha + \beta = 1, \qquad \alpha\beta = -1, \qquad \alpha - \beta = \sqrt{5}.$$

## 6. Cesaro's Observation

For our first use of Binet's formula we consider a remarkable observation by the Italian mathematician E. Cesaro (1859–1906).

Expanding $(1 + 2x)^n$ by the binomial theorem, suppose we get

$$(1 + 2x)^n = a_0 + a_1x + a_2x^2 + \cdots + a_rx^r + \cdots + a_nx^n.$$

Now, instead of the powers of $x$, let us substitute the corresponding Fibonacci numbers, that is, $f_r$ for $x^r$. This gives the following expression $S$:

$$S = a_0f_0 + a_1f_1 + a_2f_2 + \cdots + a_nf_n.$$

Is it not amazing that, for every $n$, the value of $S$ turns out to be the $3n$th Fibonacci number $f_{3n}$? The proof is short and sweet.

*Proof.* By the binomial theorem we have

$$(1 + 2x)^n = \sum_{k=0}^{n} \binom{n}{k} 2^k x^k.$$

Thus

$$S = \sum_{k=0}^{n} \binom{n}{k} 2^k f_k = \sum_{k=0}^{n} \binom{n}{k} 2^k \left( \frac{\alpha^k - \beta^k}{\alpha - \beta} \right)$$

$$= \frac{1}{\alpha - \beta} \left[ \sum_{k=0}^{n} \binom{n}{k} 2^k \alpha^k - \sum_{k=0}^{n} \binom{n}{k} 2^k \beta^k \right]$$

$$= \frac{1}{\alpha - \beta} [(1 + 2\alpha)^n - (1 + 2\beta)^n].$$

However, since $\alpha^2 = \alpha + 1$, we have

$$1 + 2\alpha = \alpha^2 + \alpha = \alpha(1 + \alpha)$$

$$= \alpha(\alpha^2) = \alpha^3.$$

Similarly $1 + 2\beta = \beta^3$. Therefore

$$S = \frac{1}{\alpha - \beta} [(\alpha^3)^n - (\beta^3)^n] = \frac{\alpha^{3n} - \beta^{3n}}{\alpha - \beta},$$

which is just Binet's formula for $f_{3n}$.

**Exercise**

Prove that $S = \Sigma_{k=0}^{n} \binom{n}{k} f_k = f_{2n}$ (another Cesaro observation).

## 7. The Lucas Sequence

The Fibonacci sequence is certainly interesting enough on its own. However, it is twice as exciting when considered in conjunction with a companion sequence which was made famous by the French number theorist Edward Lucas (1842–1891).

Since any characteristic property may be used to define a sequence, there are many ways we might choose to introduce the Lucas numbers. In the present case let us say that the $n$th term $L_n$ in the Lucas sequence is given by

$$L_n = f_{n-1} + f_{n+1}.$$

Thus we have $L_1 = 1$, $L_2 = 3$, $L_3 = 4$, and so on.

$\{f_n\}$: 0, 1, 1, 2, 3, 5, 8, 13, 21, 34, 55, 89, 144, ...

$\{L_n\}$: 1, 3, 4, 7, 11, 18, 29, 47, 76, 123, 199, ... .

Since the Fibonacci numbers are connected by the fundamental recursion

$$f_n = f_{n-1} + f_{n-2},$$

it is immediate that the Lucas numbers are likewise related: for $n > 2$,

$$L_n = L_{n-1} + L_{n-2}.$$

As in the case of the Fibonacci sequence, a value for $L_0$ is often desirable. Using $L_2 = L_1 + L_0$, we adopt $L_0 = 2$.

We note that a Lucas number is always bigger than the corresponding Fibonacci number except for $L_1$ and $f_1$ which equal 1.

Now let us use Binet's formula to derive a similar formula for $L_n$. We have

$$L_n = f_{n-1} + f_{n+1} = \frac{\alpha^{n-1} - \beta^{n-1}}{\alpha - \beta} + \frac{\alpha^{n+1} - \beta^{n+1}}{\alpha - \beta}$$

$$= \frac{1}{\alpha - \beta}\left[\alpha^n\left(\frac{1}{\alpha} + \alpha\right) - \beta^n\left(\frac{1}{\beta} + \beta\right)\right].$$

Now, substituting $(1 + \sqrt{5})/2$ for $\alpha$, we quickly see that $(1/\alpha) + \alpha$ reduces to $\sqrt{5}$, which is merely $\alpha - \beta$. Similarly, $(1/\beta) + \beta = -\sqrt{5} = -(\alpha - \beta)$. Therefore the desired Binet formula for the Lucas numbers is simply

$$L_n = \alpha^n + \beta^n.$$

An immediate consequence of this is the extension of the above observations of Cesaro to the Lucas numbers:

$$\text{(i)} \quad S = \sum_{k=0}^{n} \binom{n}{k} 2^k L_k = L_{3n} \qquad \text{(ii)} \quad S = \sum_{k=0}^{n} \binom{n}{k} L_k = L_{2n}.$$

Another nice result is the compact factoring $f_{2n} = f_n L_n$:

$$f_{2n} = \frac{\alpha^{2n} - \beta^{2n}}{\alpha - \beta} = \left(\frac{\alpha^n - \beta^n}{\alpha - \beta}\right)(\alpha^n + \beta^n) = f_n L_n.$$

A more complicated result, almost as easy to prove, is

$$f_{m+p} + (-1)^{p+1} f_{m-p} = f_p L_m.$$

Since we shall have occasion to put this to good use later on, let us spend a minute now to establish it. We would like to show that

$$\frac{\alpha^{m+p} - \beta^{m+p}}{\alpha - \beta} + (-1)^{p+1} \frac{\alpha^{m-p} - \beta^{m-p}}{\alpha - \beta} = \frac{\alpha^p - \beta^p}{\alpha - \beta}(\alpha^m + \beta^m)$$

that is,

$$\alpha^{m+p} - \beta^{m+p} + (-1)^{p+1}(\alpha^{m-p} - \beta^{m-p}) = \alpha^{m+p} + \alpha^p\beta^m - \alpha^m\beta^p - \beta^{m+p}$$

or

$$(-1)^{p+1}(\alpha^{m-p} - \beta^{m-p}) = \alpha^p\beta^m - \alpha^m\beta^p.$$

Recalling that $\alpha\beta = -1$, the left-hand side is

$$\begin{aligned}(-1)(\alpha\beta)^p(\alpha^{m-p} - \beta^{m-p}) &\\ = -(\alpha^m\beta^p - \alpha^p\beta^m) &\\ = \alpha^p\beta^m - \alpha^m\beta^p, &\end{aligned}$$

as required.

We do not have to delay using this fact, for it quickly leads to a proof of the interesting result:

> *The sum of any $4n$ consecutive Fibonacci numbers is divisible by $f_{2n}$.* For example, the sum of any 20 consecutive Fibonacci numbers is divisible by $55 = f_{10}$.

*Proof.*

$$\begin{aligned} S = \sum_{k=a+1}^{a+4n} f_k &= s_{a+4n} - s_a \\ &= (f_{a+4n+2} - 1) - (f_{a+2} - 1) \\ &= f_{a+4n+2} - f_{a+2}. \end{aligned}$$

Putting $m = a + 2n + 2$ and $p = 2n$ in the above equation, we get

$$S = f_{a+4n+2} - f_{a+2} = f_{2n} L_{a+2n+2},$$

showing the desired divisibility.

## 8. A Favorite Property

Now let us consider a property of the Lucas numbers which I find particularly delightful.

THEOREM. *Suppose p is a prime number* $> 3$, *and that* $p^k$ *is any (positive integral) power of it. Then the* $2p^k$*th Lucas number* $L_{2p^k}$ *always ends in a* 3.

*Proof.* Numbers of the form $6m$, $6m + 2$, $6m + 3$, and $6m + 4$ are obviously composite for $m > 0$. Thus a prime $p > 3$ is either of the form $6m + 1$ or $6m - 1$, that is,

$$p \equiv \pm 1 \pmod 6.$$

Accordingly,

$$p^k \equiv \pm 1 \pmod 6,$$

and we have $p^k = 6m \pm 1$ for some integer $m$ and the appropriate choice of sign. Thus $2p^k = 12m \pm 2$.

Now the Lucas numbers end in the digits

$$1, 3, 4, 7, 1, 8, 9, 7, 6, 3, 9, 2, 1, 3, 4, 7, \ldots$$

(since $L_n = L_{n-1} + L_{n-2}$, we need only know the last digits of the two previous terms in order to determine the last digit of $L_n$). It is plain that this sequence repeats with a period of length 12:

$$1, 3, 4, 7, 1, 8, 9, 7, 6, 3, 9, 2.$$

Since the second and tenth numbers in the period are 3's, counting along the sequence to the $(12m \pm 2)$th will always bring one to a 3.

## 9. An Unlikely Property

Another rather unexpected property of the Lucas numbers concerns the complex unit $i = \sqrt{-1}$. For $n > 1$, the $n \times n$ determinant

$$D_n = \begin{vmatrix} 3 & i & 0 & 0 & \cdots & 0 \\ i & 1 & i & 0 & \cdots & 0 \\ 0 & i & 1 & i & \cdots & 0 \\ 0 & 0 & i & 1 & \cdots & 0 \\ \cdots & \cdots & \cdots & \cdots & \cdots & \cdots \\ 0 & 0 & 0 & 0 & \cdots \; 1 & i \\ 0 & 0 & 0 & 0 & \cdots \; i & 1 \end{vmatrix} = L_{n+1}.$$

Easy calculations yield $D_1 = 3$, $D_2 = 4$, and by an appropriate expansion of the determinant (a nice exercise), we get in general that

$$D_n = D_{n-1} + D_{n-2}.$$

## 10. A Most Remarkable Connection

Many properties of the Fibonacci sequence are shared by the Lucas numbers. For example, the geometric argument that we applied to the sum of the squares of the Fibonacci numbers also yields

$$\sum_{k=0}^{n} L_k^2 = L_n L_{n+1} - 2$$

(two additional unit squares need to be introduced at the beginning to make the first two Lucas squares abut, and they need to be deducted at the end). In the present section, however, we will establish one of the most remarkable of all interconnections between the two sequences.

As $x$ runs through the positive integers, every now and again the number $5x^2 + 4$ is a perfect square. The same is true of $5x^2 - 4$. When $x = 1$, both these quantities are squares. However, since 1 and 9 are the only squares which differ by 8, never again are both $5x^2 + 4$ and $5x^2 - 4$ perfect squares for the same value of $x$.

It turns out that one of $5x^2 + 4$, $5x^2 - 4$ is a perfect square every time $x$ is equal to a Fibonacci number. And $x = f_n$ marks the only occasions that this is so. That is to say, either

$$5x^2 + 4 = y^2 \qquad \text{or} \qquad 5x^2 - 4 = y^2$$

has a solution $(x, y)$ in positive integers if and only if $x$ is a Fibonacci number. Now this is certainly exciting enough. However, considering the times when one of these functions actually is a perfect square $y^2$, would you care to venture a guess at the corresponding value of $y$? That's right, $y$ is always the Lucas number $L_n$ which corresponds to the Fibonacci number $x = f_n$. Thus we have the amazing theorem

THEOREM. *Either* $5x^2 + 4 = y^2$ *or* $5x^2 - 4 = y^2$ *has a solution* $(x, y)$ *in positive integers if and only if, for some* $n$, $(x, y) = (f_n, L_n)$.

*Proof.* (a) Suppose, for some $n$, $(x, y) = (f_n, L_n)$. Using the Binet formulas it is a simple matter to verify the relation

$$5f_n^2 + 4(-1)^n = L_n^2,$$

that is, either

$$5f_n^2 + 4 = L_n^2 \quad \text{or} \quad 5f_n^2 - 4 = L_n^2$$

for every value of $n$. Thus the condition is sufficient.

(b) Suppose that $(x, y)$ satisfies either $5x^2 + 4 = y^2$ or $5x^2 - 4 = y^2$. Since we never need to specify which of the two signs $\pm$ is being considered, let us adopt the concise form $5x^2 \pm 4 = y^2$ for "either $5x^2 + 4 = y^2$ or $5x^2 - 4 = y^2$."

It is an easy calculation to check that, for $x = 1, 2, 3$, the only solutions to $5x^2 \pm 4 = y^2$ are (1, 1), (1, 3), (2, 4), and (3, 7), which are in fact the pairs $(f_n, L_n)$ for $n = 1, 2, 3, 4$. Every other solution $(x, y)$, then, must have $x \geq 4$.

Let us review the entire list of solutions $(x, y)$ of $5x^2 \pm 4 = y^2$ to see whether or not each of them is a pair of the type $(f_n, L_n)$. Let us take them in order of increasing $x$. For $x = 1, 2, 3$ we have seen already that all the solutions $(x, y)$ are indeed examples of $(f_n, L_n)$. Therefore, if a solution $(x, y)$ ever *fails* to be an $(f_n, L_n)$, it must happen for a *first* time as we check through the $(x, y)$ corresponding to $x = 4, 5, 6, \ldots$.

We proceed indirectly. Suppose the first failure to be encountered is $(\bar{x}, \bar{y})$ (if there is more than one failure with this minimum value $\bar{x}$ of $x$, pick any one of them for $(\bar{x}, \bar{y})$). Thus every solution which has $x < \bar{x}$ is *not* of the failing variety. From our initial calculations, we have that $\bar{x} \geq 4$.

It is not difficult to see that if $x$ is a Fibonacci number $f_n$, then the value of $y$ will have to be the corresponding Lucas number $L_n$. To this end, suppose that $(f_n, y)$ is a solution. Then

$$5f_n^2 \pm 4 = y^2.$$

Now, in part (a) above, we used the identity

$$5f_n^2 + 4(-1)^n = L_n^2.$$

Combining these equations, we get

$$5f_n^2 = y^2 \pm 4 = L_n^2 - 4(-1)^n,$$

giving

$$y^2 - L_n^2 = \pm 4 - 4(-1)^n.$$

The right-hand side here has only three possible values, namely, 0, 8, or $-8$. Consequently, the squares $y^2$ and $L_n^2$ must either be equal or differ by 8. Thus, in order to escape from equality we would have to have $y^2$ and $L_n^2$ take the values 1 and 9, making $y$ either 1 or 3. But for $y = 1$ or 3, we easily calculate that $x$ must be 1, yielding the solutions (1, 1) and (1, 3). Since both these solutions *are* examples of $(f_n, L_n)$, we still have not escaped from $y = L_n$. That is to say, there is no escaping $y = L_n$ when $x = f_n$.

In order to have a solution $(x, y)$ fail to be an $(f_n, L_n)$, then, it must fail to have $x = f_n$. In particular, it must be that, for the failing solution $(\bar{x}, \bar{y})$, we have $\bar{x} \neq$ a Fibonacci number.

Now, because $\bar{x} \geq 4$, we shall see that $\bar{y}$ is subject to the restriction $2\bar{x} < \bar{y} < 3\bar{x}$. If $\bar{y}$ were to violate this restriction, the value of $\bar{y}^2$ would turn out either to be too big or too small to equal either of the quantities $5x^2 \pm 4$. For example, suppose $\bar{y} \leq 2\bar{x}$. In this case, we would have $\bar{y}^2 \leq 4\bar{x}^2$, making it clearly too small to equal $5x^2 + 4$, and in order to have it equal to $5x^2 - 4$, we would have to have

$$5\bar{x}^2 - 4 = \bar{y}^2 \leq 4\bar{x}^2,$$

and

$$\bar{x}^2 \leq 4, \quad \text{making} \quad \bar{x} \leq 2.$$

But we know already that $\bar{x} \geq 4$, and we would have a contradiction. A similar contradiction follows the assumption $\bar{y} \geq 3\bar{x}$.

Because $(\bar{x}, \bar{y})$ is a solution, we have

$$5\bar{x}^2 \pm 4 = \bar{y}^2,$$

which shows that $\bar{x}$ and $\bar{y}$ must have the same parity. Consequently, $\bar{y} - \bar{x}$ must be some even integer $2t$, and we have

$$\bar{y} = \bar{x} + 2t \quad (\text{clearly } \bar{y} > \bar{x}).$$

Since $2\bar{x} < \bar{y} < 3\bar{x}$, we have $2\bar{x} < \bar{x} + 2t < 3\bar{x}$, yielding

$$\frac{\bar{x}}{2} < t < \bar{x}.$$

Substituting for $\bar{y}$ in $5\bar{x}^2 \pm 4 = \bar{y}^2$, we get

$$5\bar{x}^2 \pm 4 = \bar{x}^2 + 4t\bar{x} + 4t^2,$$

$$4\bar{x}^2 - 4t\bar{x} - 4t^2 \pm 4 = 0,$$

$$\bar{x}^2 - t\bar{x} - (t^2 \pm 1) = 0.$$

Solving for $\bar{x}$, we get

$$\bar{x} = \frac{t \pm \sqrt{t^2 + 4(t^2 \pm 1)}}{2},$$

giving

$$2\bar{x} = t \pm \sqrt{5t^2 \pm 4}.$$

Thus, transposing and squaring yields

$$5t^2 \pm 4 = (2\bar{x} - t)^2 = s^2, \quad \text{for some integer } s.$$

Because $\bar{x}/2 < t < \bar{x}$, it follows that $s = 2\bar{x} - t$ is a positive integer $(> \bar{x})$. We have found, then, a solution $(t, s)$ in positive integers of our equations $5x^2 \pm 4 = y^2$. Having noted already that $t < \bar{x}$, which is the minimum value of $x$ for a solution which fails to be an $(f_n, L_n)$, we conclude that this particular solution $(t, s)$ cannot be of the failing variety and that, for some $n$, we have

$$t = f_n \quad \text{and} \quad s = L_n.$$

Going back to the equation $2\bar{x} = t \pm \sqrt{5t^2 \pm 4}$, which is

$$2\bar{x} = t \pm s,$$

we obtain

$$2\bar{x} = f_n \pm L_n.$$

Since $L_n = f_{n-1} + f_{n+1}$, we have $L_n > f_n$ except for the single case of $L_1 = f_1 = 1$. Because $\bar{x} \geq 4$, it can never be true that

$$2\bar{x} = f_n - L_n.$$

Therefore, it follows that

$$\begin{aligned} 2\bar{x} &= f_n + L_n \\ &= f_n + f_{n-1} + f_{n+1} \\ &= f_{n+1} + f_{n+1}, \end{aligned}$$

giving the contradiction

$$\bar{x} = f_{n+1},$$

a Fibonacci number. This completes the proof.

In passing we note the kindred result for Lucas numbers:

> *$x$ is a Lucas number if and only if either $5x^2 + 20$ or $5x^2 - 20$ is a perfect square.*

## 11. Another Remarkable Connection

Now we consider a startling result of an entirely different kind:

$$e^{L_1x+(L_2/2)x^2+(L_3/3)x^3+\cdots} = f_1 + f_2x + f_3x^2 + \cdots.$$

The proof consists of a neat combination of elementary generating functions and freshman calculus.

We begin by deriving a generating function for the Fibonacci sequence. Let the Fibonacci numbers be the coefficients of a power series as shown:

$$F(x) = f_1 + f_2x + f_3x^2 + \cdots + f_{n+1}x^n + \cdots.$$

Then

$$x \cdot F(x) = f_1x + f_2x^2 + \cdots + f_nx + \cdots,$$

and

$$x^2 \cdot F(x) = f_1x^2 + \cdots + f_{n-1}x^n + \cdots.$$

Subtracting the second and third lines from the first gives

$$(1 - x - x^2)F(x) = f_1 + (f_2 - f_1)x = 1,$$

and

$$F(x) = (1 - x - x^2)^{-1}.$$

Now consider the series

$$L(x) = L_1 x + \frac{L_2}{2}x^2 + \frac{L_3}{3}x^3 + \cdots + \frac{L_n}{n}x^n + \cdots.$$

Using Binet's formula and the power series for $\log(1 - y)$, we have

$$\begin{aligned}
L(x) &= (\alpha + \beta)x + \frac{\alpha^2 + \beta^2}{2}x^2 + \frac{\alpha^3 + \beta^3}{3}x^3 + \cdots \\
&= \left[(\alpha x) + \frac{(\alpha x)^2}{2} + \frac{(\alpha x)^3}{3} + \cdots\right] \\
&\quad + \left[(\beta x) + \frac{(\beta x)^2}{2} + \frac{(\beta x)^3}{3} + \cdots\right] \\
&= -\log(1 - \alpha x) - \log(1 - \beta x) \\
&= -\log[(1 - \alpha x)(1 - \beta x)] \\
&= -\log[1 - (\alpha + \beta)x + \alpha\beta x^2] \\
&= -\log(1 - x - x^2) \qquad (\text{recall } \alpha + \beta = 1,\ \alpha\beta = -1) \\
&= \log(1 - x - x^2)^{-1} \\
&= \log F(x),
\end{aligned}$$

and we have

$$e^{L(x)} = F(x),$$

as desired.

## 12. Some Problems in Combinatorics and Probability

In this section we turn to a handful of delightful problems in combinatorics and probability.

(i) In how many ways can a set of $n$ dominos be arranged to cover a $2 \times n$ checkerboard?

A $2 \times n$ board is simply two rows of $n$ squares. There isn't a great deal of leeway in placing a domino on the board, for it must go either "up and down" or "across" the board. Nevertheless there are many ways to cover a board of any appreciable length, say $2 \times 10$.

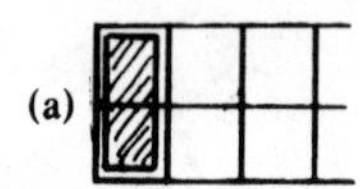


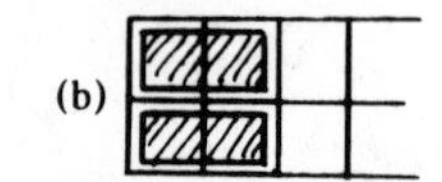


Let $a_n$ denote the required number of arrangements. Now an arrangement can begin at the left side in only two ways—(a) a single vertical domino, or (b) two horizontal dominos. In order to complete an arrangement of type (a), we need to cover the remaining $2 \times (n-1)$ board with $n-1$ dominos. By definition, there are $a_{n-1}$ ways of doing this, and so there must be $a_{n-1}$ arrangements of type (a). Similarly, the number of arrangements of type (b) is $a_{n-2}$. Altogether, then, we have

$$a_n = a_{n-1} + a_{n-2}.$$

From the early trivial cases we find that $a_1 = 1$ and $a_2 = 2$, implying that the sequence $\{a_i\}$ proceeds

$$1, 2, 3, 5, 8, 13, 21, \ldots,$$

having the general term $a_n = f_{n+1}$, the $(n+1)$th Fibonacci number.

If we had not traded on our knowledge of the Fibonacci numbers, this problem might have been much more difficult, for from Binet's formula we see that one way of stating the numerical answer is

$$a_n = \frac{1}{\sqrt{5}}\left[\left(\frac{1+\sqrt{5}}{2}\right)^{n+1} - \left(\frac{1-\sqrt{5}}{2}\right)^{n+1}\right],$$

a value which might not be easy to come upon in simpler form.

(ii) In how many ways can one make a selection from the first $n$ positive integers $(1, 2, 3, \ldots, n)$ without taking two consecutive numbers (counting the empty set as a selection)?

Let $a_n$ denote the required number. Now, a selection either contains the number $n$ or it doesn't. If it doesn't, then it must be one of the $a_{n-1}$ acceptable selections that are possible from the first $n - 1$ positive integers $(1, 2, \ldots, n - 1)$. If it does contain $n$, then it must avoid the consecutive value $n - 1$, and the *remainder* of the selection must constitute one of the $a_{n-2}$ ways of making such a selection from the first $n - 2$ integers $(1, 2, \ldots, n - 2)$. We must note that there may not be any other numbers in the selection, for $n$ constitutes a selection by itself. In this case, the remainder of the selection is the empty set. Since the empty set *is* counted among our $a_{n-2}$ remainders, we have that the number of selections containing $n$ is

(number of ways of picking $n$)(number of ways to complete the selection)

$$= (1)(a_{n-2}) = a_{n-2}.$$

Consequently, we again have that $a_n = a_{n-1} + a_{n-2}$. Because the empty set is counted, we have $a_1 = 2$ and $a_2 = 3$. Therefore the sequence $\{a_i\}$ proceeds 2, 3, 5, 8, 13, ..., with general term $a_n = f_{n+2}$, the $(n + 2)$th Fibonacci number.

(iii) What is the probability, in $n$ tosses of a fair coin, of not getting two heads in a row?

Many teachers of mathematics seem to take uncommon pleasure in posing the same problem in umpteen different disguises. What a delightful surprise it is to realize that, beneath a thin camouflage, the present problem is nothing more than the one we just solved.

A row of $n$ $H$'s and $T$'s, representing the outcome of the experiment, is acceptable if and only if no two $H$'s are consecutive. In an acceptable outcome, then, the set of numbers which mark the positions occupied by the $H$'s constitute a selection from $(1, 2, \ldots, n)$ which contains no consecutive numbers, and conversely. Thus there are $a_n = f_{n+2}$ successful outcomes, yielding a probability of

$$\frac{f_{n+2}}{2^n}.$$

And now for the highlight of this section.

> (iv) How many selections can be made from $(1, 2, \ldots, n)$, *placed around a circle*, without taking two consecutive numbers (counting the empty set)?

Arranging the numbers around a circle introduces the complication that 1 and $n$ are consecutive. However, let us try to proceed as before. Let $b_n$ denote the required number and consider the selections to be separated into two groups according to whether or not they contain the number $n$.

If a selection does not contain $n$, then, imagining the other numbers around the circle to have been straightened out along a line, we see that the selection is simply one of the acceptable selections which can be taken from the *row* $(1, 2, \ldots, n - 1)$. Using the same notation as in problem (ii), we have that the number of selections which do not contain $n$ is $a_{n-1}$.

Suppose that a selection does contain $n$. Then both its neighbors 1 and $n - 1$ must be avoided. The rest of the selection, then, constitutes a selection from the *row* of numbers $(2, 3, \ldots, n - 2)$. Of course, by subtracting 1 from each number that is chosen, a selection from $(2, 3, \ldots, n - 2)$ becomes one of the $a_{n-3}$ selections from $(1, 2, \ldots, n - 3)$, and conversely. Thus there are $a_{n-3}$ ways of making the selection if $n$ is chosen. Altogether, then, we have

$$b_n = a_{n-1} + a_{n-3}.$$

Recalling that $a_n = f_{n+2}$, we have

$$\begin{aligned} b_n &= f_{n+1} + f_{n-1} \\ &= L_n, \end{aligned}$$

the $n$th Lucas number! We close the section with a couple of exercises that you might enjoy.

**Exercises**

1. The number 4 can be expressed as an ordered sum of 1's and 2's in 5 ways:

$$1 + 1 + 1 + 1, \quad 1 + 1 + 2, \quad 1 + 2 + 1, \quad 2 + 1 + 1, \quad 2 + 2.$$

In how many ways can the positive integer $n$ be expressed as an ordered sum of 1's and 2's? (also, in how many ways as an unordered sum of 1's and 2's?)

2. In how many ways can $n$ tosses of a fair coin yield an outcome which has no three consecutive results the same, that is, having neither three consecutive heads nor three consecutive tails? (*The Fibonacci Quarterly*, 1969, pp. 222–223)

## 13. Representations

Now let us embark along an entirely different path to consider some topics in the fascinating area of "representations".

(a) Let the Fibonacci sequence be denoted by $F$:

$$F = (1, 1, 2, 3, 5, 8, 13, 21, 34, 55, 89, 144, \ldots).$$

Suppose we pass along the sequence and see whether we can pick out a set of terms which add up to 100. It doesn't take long to find one; for example, $89 + 8 + 3 = 100$. It turns out that this can be done for any positive integer that might be specified. Because of this, $F$ is said to be *complete*. In passing along a sequence we eventually encounter every term, and so the order of the terms has no bearing on the problem of selecting a subsequence to have a prescribed sum. By the same token, it does no harm to arrange the terms in order of magnitude, and such a convention does make it easier for us to discuss the subject. Suppose, then, that $V = (v_1, v_2, v_3, \ldots)$ is a sequence of positive integers arranged in nondecreasing order. We define $V$ to be *complete* if every positive integer $n$ is the sum of some subsequence of $V$, that is,

$$n = \sum_{i=1}^{\infty} a_i v_i, \qquad \text{where } a_i = 0 \text{ or } 1.$$

We shall soon prove that $F$ is complete. In fact, we shall see that $F$ is complete even when any one of its terms is deleted. Dropping any two terms, however, does succeed in destroying its completeness.

Fortunately there is a simple, central theorem in this area. It was published first in the *American Mathematical Monthly* (1961, pp.

557–560) by John L. Brown, Jr. (Pennsylvania State University). It establishes a very useful criterion for completeness.

Obviously the first, and smallest, term $v_1$ must be 1 in order to provide a subsequence with sum equal to 1. In trying to find a subsequence with sum $n$, we are obliged to forsake every term $v_k$ that exceeds $n$. Of course, for $n = v_k$ we can simply take the subsequence $v_k$ itself. In a sense, the most difficult sums to produce are those which immediately precede the terms $v_k$, that is, the numbers

$$n = v_k - 1.$$

In producing the number $v_k - 1$ we can only use the terms $v_1, v_2, \ldots, v_{k-1}$, since all the remaining terms are too big. If it should ever happen that even *all* these available terms do not suffice to give a sum as big as $v_k - 1$, there is clearly no way of producing the number $v_k - 1$. Thus, for completeness, it is certainly *necessary* that the sum of the first $k - 1$ terms carries us as far as the number $v_k - 1$. Brown's theorem claims essentially that this condition is also sufficient.

BROWN'S CRITERION. *V is complete if and only if*

(i) $v_1 = 1$, and
(ii) for all $k = 2, 3, \ldots,$

$$s_{k-1} = v_1 + v_2 + \cdots + v_{k-1} \geq v_k - 1.$$

*Proof of the sufficiency.* We assume, then, that conditions (i) and (ii) hold, and we will show that $V$ is complete.

Now it often happens that $s_{k-1}$ is actually much bigger than the required minimum of $v_k - 1$. Therefore it is generally much more to ask that the terms $v_1, v_2, \ldots, v_{k-1}$ yield all the values of $n = 1, 2, \ldots, s_{k-1}$ than just to require that they yield the values $1, 2, \ldots, v_k - 1$. We shall see, however, that this is in fact what always happens when conditions (i) and (ii) are satisfied: the terms $v_1, v_2, \ldots, v_{k-1}$ are always sufficient to generate all the positive integers right up to their absolute limit of $s_{k-1}$.

We proceed by induction. On the strength of $s_{k-1} \geq v_k - 1$, and

$v_1 = 1$, it is a simple exercise to determine that the sequence in question must begin in one of the following ways:

$$(1, 1, 1, \ldots)$$
$$(1, 1, 2, \ldots)$$
$$(1, 1, 3, \ldots)$$
$$(1, 2, 2, \ldots)$$
$$(1, 2, 3, \ldots)$$
$$(1, 2, 4, \ldots)$$

(for example, for $k = 2$, we have $v_2 \leq s_1 + 1 = v_1 + 1 = 1 + 1 = 2$). Thus, for $k = 2$ and 3, we observe directly that the terms $v_1, v_2, \ldots, v_{k-1}$ do yield all $n$ up to $s_{k-1}$.

Suppose for some $k \geq 3$, then, that the first $k - 1$ terms $v_1, v_2, \ldots, v_{k-1}$ suffice to yield all $n$ up to $s_{k-1}$. We shall deduce from this that it follows that the first $k$ terms suffice to yield all $n$ up to $s_k$.

To this end consider the case of an arbitrary positive integer $n \leq s_k$. If $n$ is as small as $s_{k-1}$, the induction hypothesis tells us that the first $k - 1$ terms are sufficient for the task, implying that the first $k$ terms are also sufficient.

Suppose, then, that $n$ is bigger than $s_{k-1}$:

$$s_{k-1} < n \leq s_k.$$

Subtracting $v_k$ throughout, we obtain

$$s_{k-1} - v_k < n - v_k \leq s_{k-1}.$$

Now it is given that $s_{k-1} \geq v_k - 1$, i.e., $s_{k-1} - v_k \geq -1$. Hence

$$-1 < n - v_k \leq s_{k-1}, \quad \text{or} \quad 0 \leq n - v_k \leq s_{k-1}.$$

In the event that $0 = n - v_k$, we generate $n$ by simply taking the single term $v_k$ itself. Otherwise we have

$$0 < n - v_k \leq s_{k-1},$$

which makes the integer $n - v_k$ a positive integer $m \leq s_{k-1}$. According to the induction hypothesis, then, this number $m$ can be generated as a sum of a subsequence of the terms $v_1, v_2, \ldots, v_{k-1}$. Adding the term $v_k$ to this sum, we obtain

$$m + v_k = n$$

as a sum of a subsequence of $v_1, v_2, \ldots, v_k$. Thus it follows by induction that our sequence $V$ is such that, for all $k = 2, 3, \ldots$, the first $k$ terms suffice to generate all values of $n$ up to the entire sum $s_k$.

Now, because each $v_i \geq 1$, the sum $s_k \geq k$, and as $k \to \infty$, we have $s_k \to \infty$. Therefore, being able to generate all $n$ up to all $s_k$ is to be able to generate all $n = 1, 2, 3, \ldots$, and $V$ is indeed a complete sequence.

## 14. Applications

(i) *$F$ is complete.* Clearly $f_1 = 1$, and we have

$$s_{k-1} = f_1 + f_2 + \cdots + f_{k-1} = f_{k+1} - 1 \geq f_k - 1.$$

In fact, since $f_{k+1} > f_k$ in all cases except $k = 1$, the sequence $F$ actually meets Brown's requirements with something to spare.

(ii) *$F - f_r$ is complete.* Suppose some term $f_r$ is dropped from $F$. Then not both $f_1$ and $f_2$ are lost, and the first term in $F - f_r$ is still 1. Now none of the partial sums up to $f_r$ is affected by the loss of $f_r$ and therefore each satisfies Brown's criterion as in part (i). The partial sums that are in question here are of the form

$$f_1 + f_2 + \cdots + f_{r-1} + f_{r+1} + \cdots + f_m,$$

where $m > r$. For these sums we have

$$\begin{aligned} f_1 + f_2 + \cdots + f_m &= f_{m+2} - 1 - f_r \\ &\geq f_{m+2} - 1 - f_m \\ &= f_{m+1} - 1, \end{aligned}$$

as required.

We notice, however, that the very first partial sum which is affected by the omission of $f_r$ just meets Brown's criterion with nothing to spare. It is therefore ruinous to drop a second term $f_s$ as well as $f_r$: if the labels are assigned so that $s < r$, then

$$\begin{aligned} &f_1 + f_2 + \cdots + f_{s-1} + f_{s+1} + \cdots + f_{r-1} \\ &\qquad = f_{r+1} - 1 - f_s < f_{r+1} - 1, \end{aligned}$$

failing to meet Brown's requirement.

Therefore, the sequence $F - f_r - f_s$ is never complete (which is obvious in the particular case of dropping both 1's at the beginning).

(iii) COROLLARY: an alternative criterion. *If* $v_1 = 1$ *and* $v_{k+1} \leq 2v_k$, *then* $V$ *is complete.*

*Proof.* Repeatedly using $v_{k+1} \leq 2v_k$, we have

$$\begin{aligned} v_k &\leq 2v_{k-1} \\ &= v_{k-1} + v_{k-1} \\ &\leq v_{k-1} + 2v_{k-2} \\ &= v_{k-1} + v_{k-2} + v_{k-2} \\ &\leq v_{k-1} + v_{k-2} + 2v_{k-3} \\ &\cdots\cdots\cdots\cdots\cdots\cdots \\ &\leq v_{k-1} + v_{k-2} + \cdots + 2v_1. \end{aligned}$$

Since $v_1 = 1$, we have

$$v_k - 1 \leq v_{k-1} + v_{k-2} + \cdots + v_1,$$

satisfying Brown's criterion and making $V$ complete.

*An Application.* In view of this alternative criterion, the following result concerning the elusive prime numbers is an immediate consequence of the well-known theorem called Bertrand's Postulate, which states that, for $n > 1$, there always exists a prime number $p$ between $n$ and $2n$. Let $V$ be the sequence of primes following an initial term of 1:

$$V = (1, 2, 3, 5, 7, 11, 13, 17, 23, 29, 31, \ldots).$$

In general, then, $v_k = p_{k-1}$, the $(k - 1)$th prime number. By Bertrand's Postulate there exists a prime $p$ between $p_k$ and $2p_k$. Now, if any prime occurs in this range, then certainly the "very next" prime $p_{k+1}$ must do so, and we have

$$p_k < p_{k+1} < 2p_k.$$

Since $2 \leq 2 \cdot 1$ for the first two terms of $V$, we have in all cases that

$$v_{k+1} \leq 2v_k,$$

implying the remarkable result that the sequence of primes, headed by an initial term 1, which is renowned for its multiplicative completeness, also possesses the additive completeness of the present discussion.

In fact, it is known that, for $n \geq 6$, there are always two prime numbers between $n$ and $2n$. Therefore, for $k \geq 4$, we have

$$p_k < p_{k+1} < p_{k+2} < 2p_k.$$

Consequently, the deletion of a single prime $> 7$ from our sequence $V$ is not enough to negate the property

$$v_{k+1} \leq 2v_k,$$

implying that $V$ is complete even with any prime $> 7$ deleted. But, to our astonishment, we see that our key property remains valid, implying the completeness of $V$, no matter how many primes $> 7$ might be deleted, provided only that no two consecutive primes are dropped.

(iv) *Weakly Complete Sequences.* A sequence $V$, although unable to produce some numbers at the beginning, might be able to generate all numbers beyond some point $N$. Such sequences, commanding an infinitely-long range of successes, do possess a large measure of completeness, and we shall say that they are "weakly" complete in contrast to the "strongly" complete sequences which are capable of generating all positive integers. We have seen that $F$ and $F - f_r$ are strongly complete, while $F - f_r - f_s$ is not. Now, what do you think: is $F - f_r - f_s$ good enough to be weakly complete? As we shall see, dropping the second term $f_s$ spoils everything.

THEOREM. *$F - f_r - f_s$ is not even weakly complete.*

*Proof.* Suppose $s < r$. We saw earlier that the number $f_{r+1} - 1$ was unattainable as a sum of a subsequence of $F - f_r - f_s$. We will use this result as the basis of an induction that the number $n = f_{r+1+2t} - 1$ is unattainable for all $t = 0, 1, 2, \ldots$.

The result is established, then, for $t = 0$. Suppose, for some value $t \geq 0$, that the number $n = f_{r+1+2t} - 1$ is unattainable as a sum of a subsequence of $F - f_r - f_s$. Consider now the number $n = f_{r+1+2(t+1)} - 1$, that is,

$$n = f_{(r+1+2t)+2} - 1.$$

The numbers we have at our disposal are

$$f_1, f_2, \ldots, f_{s-1}, f_{s+1}, \ldots, f_{r-1}, f_{r+1}, \ldots, f_{(r+1+2t)+1}.$$

If the very last term here is not used, then the greatest sum obtainable is

$$f_1 + f_2 + \cdots + f_{r+1+2t} \qquad \text{(with } f_r, f_s \text{ missing)}$$
$$= f_{(r+1+2t)+2} - 1 - (f_r + f_s),$$

which is not enough to yield the $n$ under investigation. Thus we have no choice but to include the last term, namely, $f_{(r+1+2t)+1}$ in our subsequence. With this term in hand, the balance that is still needed is

$$(f_{(r+1+2t)+2} - 1) - f_{(r+1+2t)+1} = f_{r+1+2t} - 1,$$

which is precisely an amount that we are unable to muster. Therefore, if $f_{r+1+2t} - 1$ is unattainable, so is $f_{r+1+2(t+1)} - 1$. By induction, then, $f_{r+1+2t} - 1$ is unattainable for all $t = 0, 1, 2, \ldots$.

Since there are numbers $f_{r+1+2t} - 1$ which exceed every choice of positive integer $N$, the sequence $F - f_r - f_s$ is not even weakly complete.

It is therefore extremely surprising that the closely related sequence $V = (2, 0, 3, 2, 6, 7, 14, 20, \ldots)$, obtained from $F$ by alternately adding and subtracting unity from its terms, that is, having general term $v_n = f_n - (-1)^n$, is not only weakly complete, but remains so even *with any finite subsequence deleted.* It takes an infinite thinning to render this sequence weakly incomplete, but it does succumb to the removal of any infinity of terms. This is the subject of an excellent paper by Ron Graham (Bell Laboratories, Murray Hill, New Jersey), "A Property of Fibonacci Numbers" (*The Fibonacci Quarterly*, 1964, p. 1).

(b) *Zeckendorf's Theorem.* As we have seen, $F$ is able to meet the requirement for strong completeness with something to spare. We also saw that this slack is entirely taken up with the deletion of any one term. Now let us look briefly at another way of tightening up this surplus capacity.

$F$ is more thoroughly complete than one might suspect. Its capabilities are so great that it remains strongly complete even with a small amputation and one arm tied behind its back.

ZECKENDORF'S THEOREM (1951) (*The Fibonacci Quarterly*, 1964, pp. 163–168). *The sequence $F - 1 = (1, 2, 3, 5, 8, 13, 21, \ldots)$ is strongly complete even if one is restricted to subsequences which contain no two consecutive terms.*

However, this is absolutely as far as one can go in this direction, for, in all cases, there exists exactly one subsequence of $F - 1$ which yields the positive integer $n$.

Now this is certainly enough for most of us. But . . .

THE DUAL THEOREM (*The Fibonacci Quarterly*, 1965, pp. 1–8). *The sequence $F - 1$ is strongly complete even if one is restricted to subsequences in which no two consecutive terms are both passed over (that is, until the desired total is achieved).*

For example,

$$100 = 1 + 2 + 3 + 5 + 13 + 21 + 55.$$

(c) *Repeated Copies of V.* Consider the sequence $F^2$, having terms $v_k = f_k^2$:

$$F^2 = (1, 1, 4, 9, 25, 64, 169, \ldots).$$

Clearly $F^2$ is not complete, for there is no way to generate 3 or 7. In fact, $F^2$ is not even weakly complete. However, it is interesting to observe that two copies of $F^2$ do meet Brown's criterion for completeness:

$$2F^2 = (1, 1, 1, 1, 4, 4, 9, 9, 25, 25, 64, 64, 169, 169, \ldots).$$

It is known that $2^{n-1}$ copies of $F^n$ are complete, where the terms of $F^n$ are given by $v_k = f_k^n$.

Of course the Lucas numbers are another promising source of interesting results in the field of representations (*The Fibonacci Quarterly*, 1969, pp. 243–252). You might also enjoy the article "Primer on Representations" (*The Fibonacci Quarterly*, 1973, pp. 317–331).

**15. Divisibility**

Earlier we encountered the property that $f_n | f_m$ if $n | m$. We shall soon prove the converse (for $n \geq 3$) and, in conjunction with the

Lucas numbers, the following nice package of results has been discovered:

(a) for $n \geq 3$, $f_n | f_m$ if and only if $n$ divides $m$ (the condition $n \geq 3$ merely avoids the case of $f_2 = 1$ which divides $f_m$ whether or not $2 | m$);
(b) $L_n | f_m$ if and only if $n$ divides into $m$ an *even* number of times;
(c) $L_n | L_m$ if and only if $n$ divides into $m$ an *odd* number of times.

It turns out that the greatest common divisor of two Fibonacci numbers, $(f_m, f_n)$, is also a Fibonacci number and, to our amazement, it is the very Fibonacci number whose subscript is the greatest common divisor of the subscripts involved:

THEOREM. $(f_m, f_n) = f_{(m,n)}$.

*Proof.* (By Glen Michael, Washington State University) (*The Fibonacci Quarterly*, 1964, pp. 57–58). Let us use the notation $(f_m, f_n) = d$ and $(m, n) = c$; then we want to prove that $d = f_c$.

We know that if $n | m$, then $f_n | f_m$. Because $(m, n) = c$, we have $c | m$ and $c | n$. Thus $f_c | f_m$ and $f_c | f_n$. Consequently, $f_c$ is a common divisor of $f_m$ and $f_n$. Now the greatest common divisor of two numbers is not just the one having greatest magnitude, but it is the greatest in the sense that it contains among its factors *all* the factors that are common to the two numbers in question. As a result, the greatest common divisor of two numbers is *divisible* by every common divisor of the numbers. Accordingly, we have

$$f_c | d.$$

We shall show also that $d | f_c$, from which the desired $d = f_c$ follows.

By the Euclidean algorithm, there exist integers $a$ and $b$ such that

$$am + bn = c.$$

Since $m$, $n$, and $c$ are all positive, both $a$ and $b$ cannot be negative. On the other hand, if both are positive, then

$$c = am + bn \geq m + n > m,$$

contradicting $c \mid m$. Thus $a$ and $b$ must straddle zero, with the possibility of one of them being zero. For definiteness, suppose $a \leq 0$. Then, if we let $k = -a$, we have $k \geq 0$. In this case

$$bn = c - am = c + km.$$

Next we use the identity $f_{n+m} = f_{n-1}f_m + f_n f_{m+1}$. This gives

$$f_{bn} = f_{c+km} = f_{c-1}f_{km} + f_c f_{km+1}. \tag{1}$$

Now, we have $d \mid f_n$, and also that $f_n \mid f_{bn}$; thus $d \mid f_{bn}$; similarly $d \mid f_m$, and $f_m \mid f_{km}$; thus $d \mid f_{km}$. It follows from (1), then, that

$$d \mid f_c f_{km+1}.$$

But two consecutive Fibonacci numbers are always relatively prime (the easy proof of this is left to the reader). Thus, because $d \mid f_{km}$, it must be that $d$ is relatively prime to the next Fibonacci number, $f_{km+1}$. Consequently we have $d \mid f_c$ as desired.

The result $(f_m, f_n) = f_{(m,n)}$ goes a long way toward keeping the divisibility properties of the Fibonacci numbers in the family. This sequence is so tightfisted that, even when $f_n$ fails to divide $f_m$, it turns out that either the remainder $r$ that is produced or the complementary remainder $f_n - r$ is always a Fibonacci number. If one increases the quotient by 1, the negative remainder obtained is $-(f_n - r)$. We might say that $r$ is the least positive remainder and $f_n - r$ is the least negative remainder. One of these least remainders is always a Fibonacci number. For example,

$$\frac{f_{11}}{f_7} = \frac{89}{13}$$

has remainders 11 and $-2$, and 2 is a Fibonacci number. This general result leads to things like

$$f_n \not\equiv 4 \pmod{13},$$

because 13 is $f_7$ and neither $r = 4$ nor $13 - 4 = 9$ is a Fibonacci number.

### Exercise

Prove that no odd Fibonacci number is ever divisible by 17.

*Primes*. The question of whether or not there is an infinity of prime numbers among the Fibonacci numbers has never been answered. We can show, however, that there do exist arbitrarily long stretches of consecutive Fibonacci numbers all of which are composite. It is easy to see that the $n$ consecutive numbers

$$(n + 2)! + 3, (n + 2)! + 4, \ldots, (n + 2)! + (n + 2)$$

are all composite (3 divides the first, 4 the second, ...). Because $f_n | f_{kn}$, then, the $n$ consecutive Fibonacci numbers

$$f_{(n+2)!+3}, f_{(n+2)!+4}, \ldots, f_{(n+2)!+(n+2)}$$

are all composite ($f_3$ divides the first, $f_4$ the second, ...).

Another property that can easily be proved is the amusing

THEOREM. *No Fibonacci number* $> 8$ *ever occurs next to a prime number.*

*Proof.* By direct substitution of the Binet formulas, one can verify that

$$f_{4k} - 1 = f_{2k+1}L_{2k-1},$$
$$f_{4k+1} - 1 = f_{2k}L_{2k+1},$$
$$f_{4k+2} - 1 = f_{2k}L_{2k+2},$$
$$f_{4k+3} - 1 = f_{2k+2}L_{2k+1},$$

which show that, for all $n$, the number $f_n - 1$ is composite.

Similarly we have

$$f_{4k} + 1 = f_{2k-1}L_{2k+1},$$
$$f_{4k+1} + 1 = f_{2k+1}L_{2k},$$
$$f_{4k+2} + 1 = f_{2k+2}L_{2k},$$
$$f_{4k+3} + 1 = f_{2k+1}L_{2k+2},$$

showing that $f_n + 1$ is always composite.

*An Application of the Pigeonhole Principle.* Suppose one chooses any $n + 1$ numbers from the first $2n$ positive integers $(1, 2, \ldots, 2n)$.

By factoring all the 2's from each of these numbers, they can be expressed in the form $2^r q$, where $q$ is odd. The $n + 1$ chosen numbers thus yield $n + 1$ numbers $q$. But these odd numbers $q$ must come from the set $(1, 3, 5, \ldots, 2n - 1)$, which only has $n$ numbers in it. Therefore, by the pigeonhole principle,* some two of our numbers $q$ must be the same. Then the two chosen numbers which yield these equal values of $q$ must be such that the smaller divides the greater, for the $q$'s will cancel and the smaller power of 2 will divide the greater power. Thus we have the conclusion that, if one takes any $n + 1$ numbers from $(1, 2, \ldots, 2n)$, two will be taken such that one divides the other.

Now, since $f_n | f_m$ if $n | m$, then it follows that if one chooses any $n + 1$ of the Fibonacci numbers $(f_1, f_2, \ldots, f_{2n})$, two will be chosen such that one will divide the other. (The $n + 1$ subscripts of the chosen Fibonacci numbers must contain a pair $r$, $s$ such that $r | s$; thus $f_r | f_s$.)

## 16. Series

We conclude this long look at the Fibonacci and Lucas numbers with a couple of results on infinite series and two miscellaneous observations.

(a) Let us begin with the problem of summing the series

$$S = \sum_{n=2}^{\infty} \frac{1}{f_{n-1} f_{n+1}} = \frac{1}{1 \cdot 2} + \frac{1}{1 \cdot 3} + \frac{1}{2 \cdot 5} + \frac{1}{3 \cdot 8} + \cdots.$$

Not shrinking from a move that appears to complicate matters, we can manipulate the general term to our advantage as follows:

$$\frac{1}{f_{n-1} f_{n+1}} = \frac{f_n}{f_{n-1} f_n f_{n+1}} = \frac{f_{n+1} - f_{n-1}}{f_{n-1} f_n f_{n+1}}$$

$$= \frac{1}{f_{n-1} f_n} - \frac{1}{f_n f_{n+1}}.$$

Therefore

$$S = \left(\frac{1}{1 \cdot 1} - \frac{1}{1 \cdot 2}\right) + \left(\frac{1}{1 \cdot 2} - \frac{1}{2 \cdot 3}\right) + \left(\frac{1}{2 \cdot 3} - \frac{1}{3 \cdot 5}\right) + \cdots,$$

having partial sums

$$s_k = 1 - \frac{1}{f_{k+1}f_{k+2}}.$$

Thus

$$S = \lim_{k\to\infty} s_k = 1.$$

(b) *Millin's Series.* We shall shortly have good use for the value of

$$\lim_{n\to\infty} \frac{f_{n+r}}{f_n}.$$

Using Binet's formula, we have

$$\begin{aligned}\frac{f_{n+r}}{f_n} &= \frac{\alpha^{n+r} - \beta^{n+r}}{\alpha^n - \beta^n}\\ &= \frac{\alpha^{n+r}[1 - (\beta/\alpha)^{n+r}]}{\alpha^n[1 - (\beta/\alpha)^n]}.\end{aligned}$$

Now, $\beta = (1 - \sqrt{5})/2 = -.6$ approximately, and $\alpha = (1 + \sqrt{5})/2 = 1.6$ approximately. Thus $|\beta/\alpha| < 1$, and $\lim_{k\to\infty} (\beta/\alpha)^k = 0$. Consequently, we obtain

$$\lim_{n\to\infty} \frac{f_{n+r}}{f_n} = \alpha^r.$$

Now let us turn to the challenging problem of summing the series

$$S = \sum_{n=0}^{\infty} \frac{1}{f_{2^n}} = \frac{1}{f_1} + \frac{1}{f_2} + \frac{1}{f_4} + \frac{1}{f_8} + \cdots.$$

It was D. A. Millin, a high school student at the time, who first discovered the answer: $S = (7 - \sqrt{5})/2$.

*Proof.* As is often the case, a look at the first few partial sums reveals a telltale pattern. Let

$$S_n = \frac{1}{f_1} + \frac{1}{f_2} + \cdots + \frac{1}{f_{2^n}}$$

(actually $S_n$ is the $(n + 1)$th partial sum). Then $S = \lim_{n\to\infty} S_n$.

We have

$$S_0 = 1; \qquad S_1 = 2; \qquad S_2 = 1 + 1 + \frac{1}{3} = 2 + \frac{1}{3};$$

$$S_3 = S_2 + \frac{1}{21} = 2 + \frac{8}{21},$$

and we conjecture in general that

$$S_n = 2 + \frac{f_{2^n - 2}}{f_{2^n}}.$$

We proceed to prove this by induction. Having observed that this holds for $n = 2$ and 3 (even for $n = 1$), suppose that it is valid for some positive integer $n \geq 3$. Then we have

$$\begin{aligned} S_{n+1} = S_n + \frac{1}{f_{2^{n+1}}} &= 2 + \frac{f_{2^n-2}}{f_{2^n}} + \frac{1}{f_{2^{n+1}}} \\ &= 2 + \frac{f_{2^n-2} \cdot f_{2^{n+1}} + f_{2^n}}{f_{2^n} \cdot f_{2^{n+1}}}. \end{aligned}$$

But we know that $f_{2m}$ factors into $f_m L_m$, making

$$f_{2^{n+1}} = f_{2 \cdot 2^n} = f_{2^n} L_{2^n}.$$

This gives

$$S_{n+1} = 2 + \frac{f_{2^n-2} \cdot f_{2^n} L_{2^n} + f_{2^n}}{f_{2^n} f_{2^{n+1}}} = 2 + \frac{f_{2^n-2} L_{2^n} + 1}{f_{2^{n+1}}}.$$

Recall now the identity

$$f_{m+p} + (-1)^{p+1} f_{m-p} = f_p L_m.$$

For $m = 2^n$ and $p = 2^n - 2$, this gives

$$f_{2^n-2} L_{2^n} = f_{2^{n+1}-2} - f_2.$$

Hence,

$$\begin{aligned} S_{n+1} &= 2 + \frac{f_{2^{n+1}-2} - f_2 + 1}{f_{2^{n+1}}} \\ &= 2 + \frac{f_{2^{n+1}-2}}{f_{2^{n+1}}}. \end{aligned}$$

Therefore, by induction, the formula for $S_n$ is valid for all $n$.

Consequently we have

$$S = \lim_{n\to\infty} S_n = \lim_{n\to\infty}\left(2 + \frac{f_{2^n-2}}{f_{2^n}}\right)$$

$$= 2 + \alpha^{-2} = 2 + \frac{1}{\alpha^2} = 2 + (-\beta)^2$$

$$= 2 + \beta^2 = 2 + \frac{1 - 2\sqrt{5} + 5}{4} = \frac{7 - \sqrt{5}}{2}.$$

Had we missed the above pattern for $S_n$, we would still have had a good chance to succeed, for the series can be summed similarly by observing any of the forms

$$S_n = 1 + \frac{L_{2^n-1}}{f_{2^n}} = 2 + \frac{f_{2^n-2}}{f_{2^n}} = 3 - \frac{f_{2^n-1}}{f_{2^n}}$$

$$= 4 - \frac{f_{2^n+1}}{f_{2^n}} = 5 - \frac{f_{2^n+2}}{f_{2^n}} = 6 - \frac{L_{2^n+1}}{f_{2^n}}.$$

(c) (i) Curiously the 13th even perfect number is

$$2^{520}(2^{521} - 1),$$

and the 13th Lucas number happens to be $L_{13} = 521$.

(ii) Finally, let $t_n$ denote the $n$th triangular number,

$$t_n = 1 + 2 + 3 + \cdots + n = \frac{n(n+1)}{2}.$$

Then the number $(t_{48})^2 \cdot 5^{f_n}$, which happens to be equal to

$$5^{f_{n+96}} - 2 \cdot 5^{f_{n+48}} + 5^{f_n},$$

is always a multiple of the number

$$2^{12} \cdot 3^5 \cdot 7^3 = 341397504 \quad \text{for all } n = 1, 2, 3, \ldots.$$

**Exercises**

1. Prove that $2^n L_n$ always ends in a 2.

2. Sum the series

$$\sum_{n=1}^{\infty} \frac{f_n}{f_{n+1}f_{n+2}}.$$

3. Prove that $f_n$ is never divisible by 7 if $n$ is odd.

4. Establish geometrically that $f_n^2 = f_{n-1}^2 + 3f_{n-2}^2 + 2f_{n-2}f_{n-3}$.

5. Prove that $f_{4n}^2 + 8f_{2n}(f_{2n} + f_{6n})$ is always a perfect square.

6. Determine the value of (i) $\lim_{n\to\infty} L_n/f_n$ and (ii) $\lim_{n\to\infty} f_n^{1/n}$.

7. Prove that $f_{n+1}L_{n+2} - f_{n+2}L_n = f_{2n+1}$ and use it to show that

$$\sum_{n=1}^{\infty} \frac{f_{2n+1}}{L_n L_{n+1} L_{n+2}} = \frac{1}{3}.$$

8. Prove that

$$\sum_{k=0}^{n} \binom{n}{k} \alpha^{3k-2n} = 2^n.$$

9. If $r$ is odd, show that

$$\sum_{k=1}^{r-1} (-1)^k \binom{r}{k} f_k = 0.$$

10. How many digits are there in $f_{100}$? What is the *first* digit?

11. Determine the maximum value of $m^2 + n^2$, where $m$ and $n$ are integers satisfying $m, n \in (1, 2, \ldots, 1981)$ and $(n^2 - mn - m^2)^2 = 1$. (International Olympiad, 1981)

12. Prove that $(f_n f_{n+3}, 2f_{n+1}f_{n+2}, f_{n+1}^2 + f_{n+2}^2)$ is always a Pythagorean triple.

### References

1. V. E. Hoggatt, Jr., The Fibonacci and Lucas Numbers, Houghton-Mifflin, Boston, 1969.
2. N. N. Vorob'ev, Fibonacci Numbers, Blaisdell Publishing Co., New York, 1961.
3. The Fibonacci Quarterly, The Fibonacci Association, San Jose State University, San Jose, Ca.

CHAPTER 9

# SOME PROBLEMS IN COMBINATORICS

## 1. A Problem from Down Under

I came across this first problem as an exercise in *Combinatorial Theory: An Introduction*, by the Australians Street and Wallis [1]:

> $S = ABCD$ is a square of unit side and $n$ is an arbitrary positive integer. Inside $S$ is drawn a curve $P$, consisting only of straight segments, having length $> 2n$. Prove that some straight line $L$, which is parallel to a side of $S$, must cross $P$ at least $n + 1$ times. (The polygonal curve $P$ can be in as many pieces as you like and can even be self-intersecting.)

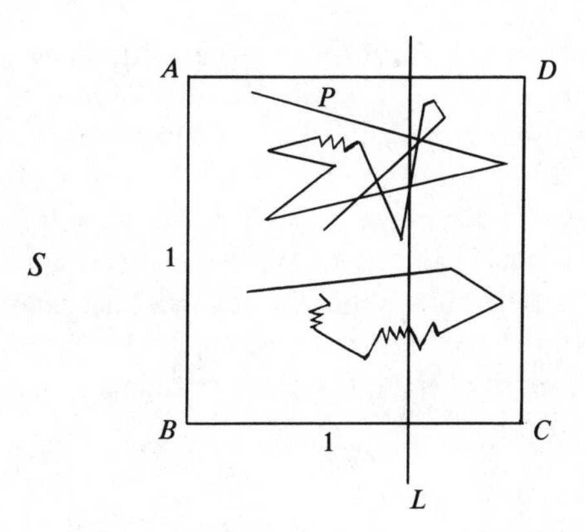


Let $EF$ be a typical segment of $P$ and let $GH$ and $IJ$ be its projections on the adjacent sides $BC$ and $AB$ of the square. From triangle $EFK$ it is clear that the sum of the projections is at least as great as the segment itself: $GH + IJ \geq EF$. Doing this for all the segments of $P$, and adding, one obtains

$$\Sigma GH + \Sigma IJ \geq \text{length of } P > 2n.$$

As a result, as least one of these sums must be $> n$; for definiteness, suppose that

$$\Sigma GH = \text{the sum of the projections on } BC > n.$$

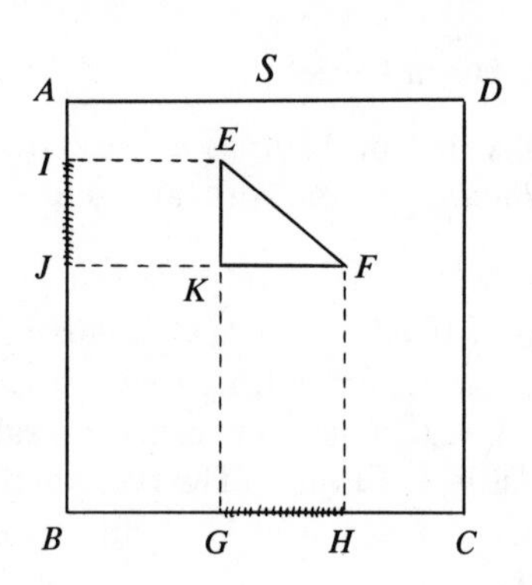


Now the entire length of $BC$ is only one unit. Consequently these projections rain down on $BC$ with enough coverage to blanket the whole stretch $n$ times and then some. Whether or not they are spread uniformly, in piling up on the base, some point $X$ of $BC$ must be covered by at least $n + 1$ of the projections—if no point of $BC$ were to lie under more than $n$ projections, then their total length would not exceed $n \cdot 1 = n$. Therefore a line $L$ through $X$, perpendicular to $BC$, would cross the $n + 1$ or more segments of $P$ whose projections cover $X$. (Another triumph of the pigeonhole principle.)

## 2. Two Problems on Partitions

Let us turn now to two more exercises from a recent text in combinatorics—*Basic Techniques of Combinatorial Theory*, by Daniel Cohen [**2**]. It is a pleasure to commend this well written exposition and treasury of good problems.

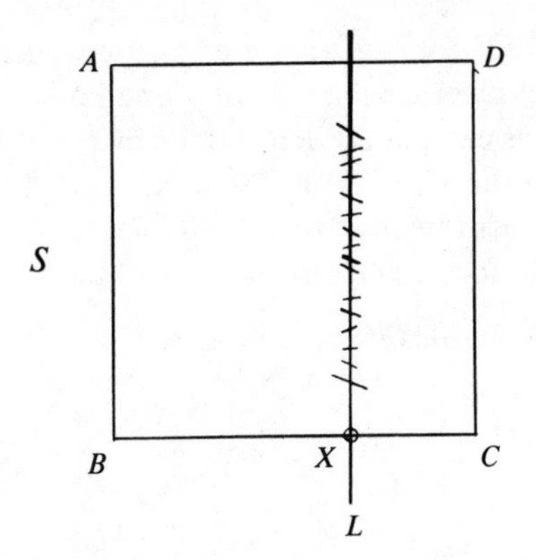


(a) *Perfect Partitions* (#64, p. 97). The 15 (unordered) partitions of 7 are

$7$
$6 + 1$
$5 + 2$
$5 + 1 + 1$
$4 + 3$
* $4 + 2 + 1$
* $4 + 1 + 1 + 1$
$3 + 3 + 1$
$3 + 2 + 2$
$3 + 2 + 1 + 1$
$3 + 1 + 1 + 1 + 1$
* $2 + 2 + 2 + 1$
$2 + 2 + 1 + 1 + 1$
$2 + 1 + 1 + 1 + 1 + 1$
* $1 + 1 + 1 + 1 + 1 + 1 + 1.$

The four partitions marked with an asterisk are special and are called "perfect". Each is not only capable of generating the number 7, but can also produce each positive integer up to 7 as a sum of a subset of its parts (which $5 + 2$, for example, is unable to do). This isn't enough to earn real distinction, for the undistinguished partition $3 + 2 + 1 + 1$ can also muster all these integers. The thing that really makes these four special is that their expressions for these integers are all *unique*: there is only one way to generate each of 1, 2, 3, 4, 5, 6 from the parts of the partition $4 + 2 + 1$, but not so from $3 + 2 + 1 + 1$; for example, the latter partition gives 5 as both $3 + 2$ and $3 + 1 + 1$. (In generating these integers, we do not distinguish between repeated parts: e.g., for the partition $2 + 2 + 2 + 1$, choosing a different 2 for use in $2 + 1$ does not destroy the uniqueness of the expression for 3.) Of course, the question we would like to ask is: How many perfect partitions are there of the positive integer $n$?

A partition of $n$ is completely determined by specifying the number of 1's it contains, the number of 2's, and so on. Suppose that the parts of a perfect partition of $n$, in order of increasing magnitude, are $a_1, a_2, \ldots, a_k$ and, for each $i$, the number of terms equal to $a_i$ is $n_i$:

$$\underbrace{(a_1 + a_1 + \cdots + a_1)}_{n_1} + \underbrace{(a_2 + a_2 + \cdots + a_2)}_{n_2} + \cdots + \underbrace{(a_k + a_k + \cdots + a_k)}_{n_k} = n.$$

Now, unless $a_1 = 1$, there would be no way to generate the integer 1. Hence we must have $a_1 = 1$, and the $n_1$ 1's provide a means of generating (uniquely) all the integers up to $n_1$. This forces the next part $a_2$ to be $n_1 + 1$, or else there would be no way to generate the integer $n_1 + 1$ ($a_2$ cannot be less than $n_1 + 1$ without spoiling the uniqueness of its representation). Therefore, $a_2 = n_1 + 1$, and with a single part equal to $a_2$ our ability to generate integers is extended to $a_2 + n_1$, which is $2a_2 - 1$, at which point a second $a_2$ comes into play (if there is one). Altogether, the $a_2$'s extend our ability to generate integers up to

$$n_1a_1 + n_2a_2 = n_1 + n_2(n_1 + 1) = (n_1 + 1)(n_2 + 1) - 1.$$

As before, unless $a_3 = (n_1 + 1)(n_2 + 1)$, there would be no way to generate this number (again, $a_3$ can't be less than this amount without compromising the uniqueness of some expression). With the $a_3$'s we can generate all the integers up to

$$\begin{aligned} &n_1a_1 + n_2a_2 + n_3a_3 \\ &\quad = [(n_1 + 1)(n_2 + 1) - 1] + n_3(n_1 + 1)(n_2 + 1) \\ &\quad = (n_1 + 1)(n_2 + 1)(n_3 + 1) - 1. \end{aligned}$$

Similarly, $a_4$ must be $(n_1 + 1)(n_2 + 1)(n_3 + 1)$, and on it goes. Eventually, reaching the $a_k$'s, we can generate all the integers up to

$$n_1a_1 + n_2a_2 + \cdots + n_ka_k = (n_1 + 1)(n_2 + 1) \cdots (n_k + 1) - 1.$$

But this final value is just the number $n$ itself, and we have

$$(n_1 + 1)(n_2 + 1) \cdots (n_k + 1) = n + 1.$$

That is to say, we obtain the "coordinates" $(n_1, n_2, \ldots, n_k)$ of a perfect partition of $n$ from each way the number $n + 1$ can be factored into a product of integers each $\geq 2$. Because the smallest part $a_1$ must always be 1 and each $a_i$ is just one more than the sum of all the smaller parts (repetitions included), these coordinates yield the complete specification of a perfect partition. Furthermore, these relations make the coordinates an ordered set, implying that there is a perfect partition for each way of ordering the factors in the product $(n + 1)$. For example, for $n = 7$ we have $n + 1 = 8$ and the following results:

| factor-ization | $n_1$ | $a_2$ | $n_2$ | $a_3$ | $n_3$ | partition |
|---|---|---|---|---|---|---|
| 8 | 7 | | | | | $1+1+1+1+1+1+1$ |
| $4 \cdot 2$ | 3 | 4 | 1 | | | $1+1+1+4$ |
| $2 \cdot 4$ | 1 | 2 | 3 | | | $1+2+2+2$ |
| $2 \cdot 2 \cdot 2$ | 1 | 2 | 1 | 4 | 1 | $1+2+4$ |

(This topic is due to Major Percy MacMahon (1854–1929), one of the leading figures in the history of combinatorics.)

(b) *A Result of Euler* (#68, p. 99). This exercise provides another opportunity to show off the versatility of generating functions.

In an earlier essay, we observed that the generating function for the unrestricted (unordered) partitions of $n$ is

$$\begin{aligned} f(x) &= (1 - x)^{-1}(1 - x^2)^{-1}(1 - x^3)^{-1} \cdots \\ &= (1 + x + x^2 + \cdots)(1 + x^2 + x^4 + \cdots) \\ &\quad \cdot (1 + x^3 + x^6 + \cdots) \cdots \\ &= \cdots + p(n)x^n + \cdots \end{aligned}$$

(the term selected from the first bracket denotes the number of 1's in the partition, and so on). This basic function can be tailored to suit various specific interests. For example, if you were interested in the number of 1's in the partitions, you could alter the first factor and use

$$f(x, y) = (1 + yx + y^2x^2 + y^3x^3 + \cdots)(1 + x^2 + x^4 + \cdots) \cdot (1 + x^3 + x^6 + \cdots) \cdots = \cdots + p(k, n)y^k x^n + \cdots,$$

where the coefficient $p(k, n)$ of $y^k x^n$ would show how many partitions of $n$ contain precisely $k$ 1's. Similarly, if you wanted to focus on partitions in which there occur exactly $k$ different values as parts, repetitions allowed, you would turn to

$$f(x, y) = (1 + yx + yx^2 + yx^3 + \cdots) \cdot (1 + yx^2 + yx^4 + yx^6 + \cdots) \cdots$$

(This looks pretty complicated, but extensive simplifications are often possible.)

The problem at hand concerns three special classes of partitions. Let $O(n)$ and $E(n)$ denote, respectively, the number of partitions of $n$ in which the number of parts is odd and even. The parts, themselves, can be either odd or even; it is the *number* of parts that is the distinction here. In the third case, let $DO(n)$ denote the number of partitions of $n$ in which the parts are all odd numbers and all different from each other; this time we aren't concerned with how many parts there are, and this represents quite a shift in character from the previous classes of partitions. For example, for $n = 6$ we have

| $O(6) = 5$ | $E(6) = 6$ | $DO(6) = 1$ |
|---|---|---|
| 6 | 5 + 1 | 5 + 1 |
| 4 + 1 + 1 | 4 + 2 | |
| 3 + 2 + 1 | 3 + 3 | |
| 2 + 2 + 2 | 3 + 1 + 1 + 1 | |
| 2 + 1 + 1 + 1 + 1 | 2 + 2 + 1 + 1 | |
| | 1 + 1 + 1 + 1 + 1 + 1 | |

The problem is to prove that $DO(n)$ is always the difference between $O(n)$ and $E(n)$:

$$|O(n) - E(n)| = DO(n).$$

Since we are interested in the total number of parts (whether odd or even) in a partition, let us use the generating function

$$\begin{aligned} f(x, y) &= (1 + yx + y^2x^2 + \cdots)(1 + yx^2 + y^2x^4 + \cdots) \\ &\quad \cdot (1 + yx^3 + y^2x^6 + \cdots) \cdots \\ &= (1 - yx)^{-1}(1 - yx^2)^{-1} \cdots (1 - yx^i)^{-1} \cdots \\ &= \sum_{n,k=0}^{\infty} a(n, k)y^k x^n. \end{aligned}$$

In this case, $a(n, k)$ denotes the number of partitions of $n$ having a total of $k$ parts. (Each part is accompanied by a $y$; the size of the part depends on the bracket from which it is taken.) Therefore, we have

$$O(n) = a(n, 1) + a(n, 3) + a(n, 5) + \cdots \quad \text{(k odd)},$$

and

$$E(n) = a(n, 2) + a(n, 4) + a(n, 6) + \cdots \quad \text{(k even)}.$$

Now the substitution $y = -1$ would change a term in $y^k x^n$ into a term in $x^n$, making negative those which have $k$ odd and positive those with $k$ even. Consequently, in the function $f(x, -1)$, the coefficient of $x^n$ is none other than $E(n) - O(n)$:

$$\begin{aligned} f(x, -1) &= \sum_{n=0}^{\infty} [E(n) - O(n)]x^n \\ &= (1 + x)^{-1}(1 + x^2)^{-1}(1 + x^3)^{-1} \cdots (1 + x^i)^{-1} \cdots \\ &= \frac{1}{1 + x} \cdot \frac{1}{1 + x^2} \cdot \frac{1}{1 + x^3} \cdots \frac{1}{1 + x^i} \cdots \\ &= \frac{1 - x}{1 - x^2} \cdot \frac{1 - x^2}{1 - x^4} \cdot \frac{1 - x^3}{1 - x^6} \cdots \frac{1 - x^i}{1 - x^{2i}} \cdots, \end{aligned}$$

giving

$$f(x, -1) = (1 - x)(1 - x^3)(1 - x^5) \cdots (1 - x^{2i+1}) \cdots,$$

displaying only odd powers of $x$.

On the other hand, in constructing a partition that is counted by $DO(n)$, one uses either one or zero l's, either one or zero 3's, and so on through the odd integers. Thus the generating function for the numbers $DO(n)$ is just

$$g(x) = (1 + x)(1 + x^3)(1 + x^5) \cdots (1 + x^{2i+1}) \cdots$$
$$= \cdots + DO(n)x^n + \cdots.$$

Since the sum of an odd number of odd integers is odd, it follows that, if $n$ is odd, the sign of every term in $x^n$ in $f(x, -1)$ will be minus (and if $n$ is even, the sign will be plus). Thus the terms which contribute to a specific $x^n$ all have the *same sign*. The final coefficients of $x^n$ in $f(x, -1)$ and $g(x)$, then, will be the same, except possibly for sign, and our argument is complete.

## 3. The Catalan Numbers and the Reflection Principle

Our final problem is one about sequences of $+1$'s and $-1$'s that was posed in 1946 by the eminent mathematicians Paul Erdös and Irving Kaplansky in *Scripta Mathematica* (pp. 73-75).

Suppose $n+1$'s and $n-1$'s are lined up in a row. There are $\binom{2n}{n}$ ways to order these $2n$ symbols since each arrangement corresponds to a way of distinguishing $n$ of the places for the $+1$'s. For example, for $n = 2$, there are 6 rows:

$$\begin{array}{cc} +1\ +1\ -1\ -1 & -1\ +1\ +1\ -1 \\ +1\ -1\ +1\ -1 & -1\ +1\ -1\ +1 \\ +1\ -1\ -1\ +1 & -1\ -1\ +1\ +1. \end{array}$$

Of course the sum of all $2n$ numbers in the row is zero. Our present interest, however, concerns the *partial sums* of the rows. While the full sum is always zero, partial sums can range as high as $+n$ and as low as $-n$. Our problem is to determine how many of the rows yield no partial sum which is negative.

Of the 6 rows for $n = 2$, only the first two escape a negative partial sum. Of the $\binom{6}{3} = 20$ rows for $n = 3$, only the 5 below have exclusively nonnegative partial sums:

$$\begin{array}{c} +1\ +1\ +1\ -1\ -1\ -1 \\ +1\ +1\ -1\ +1\ -1\ -1 \\ +1\ +1\ -1\ -1\ +1\ -1 \\ +1\ -1\ +1\ +1\ -1\ -1 \\ +1\ -1\ +1\ -1\ +1\ -1. \end{array}$$

In checking the $\binom{8}{4} = 70$ rows for $n = 4$, one finds 14 with no negative partial sums. Thus, for $n = 2, 3, 4$, respectively, we have 2 out of 6, 5 out of 20, and 14 out of 70 with the property in question. While it might slip by that 2 and 5 are divisors of 6 and 20, the 14 out of 70 makes it far less likely for the relation to go unnoticed. A look at the quotients $6/2 = 3$, $20/5 = 4$, $70/14 = 5$, points strongly to the conjecture that $1/(n + 1)$ of the rows are of the desired kind, that is $(1/(n + 1))\binom{2n}{n}$ rows.

In Daniel Cohen's book, mentioned above, the row of $+1$'s and $-1$'s is likened to a lineup of $2n$ patrons at a theater box office:

> The price of admission is 50 cents and $n$ of the people have the exact change while the other $n$ have one-dollar bills; thus each person either provides one unit of change for the cashier's later use $(+1)$ or uses up one unit of her change $(-1)$; the question is "In how many ways can the patrons be lined up so that a cashier, who begins with no change of her own, is never stuck for change?"

We shall use a very neat idea called the reflection principle to solve this problem. We can keep track of the partial sums along our row of $+1$'s and $-1$'s very nicely by plotting our progress on a coordinate plane as follows. Beginning at the origin 0, for each term absorbed into the extending sum we take a unit step in both the $x$-direction and the $y$-direction; the step in the $x$-direction is always in the positive sense while the step in the $y$-direction is in the positive sense for a $+1$ and the negative sense for a $-1$. Thus, with the inclusion of each term in the sequence, we move either upward to the right $(+1)$ or downward to the right $(-1)$.

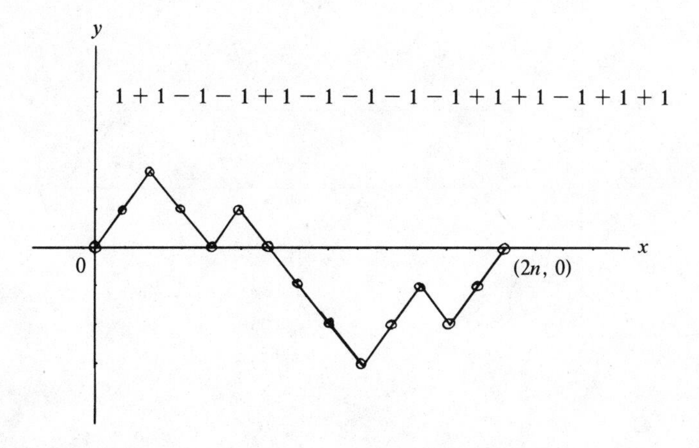

Since the total sum of the sequence is zero, we will finish our journey at the point $(2n, 0)$. Clearly the $y$-coordinate at any position represents the current value of the partial sum of the sequence. Consequently, our problem is to determine how many such zigzag paths there are from 0 to $(2n, 0)$ which do not leave the first quadrant.

Since there is a zigzag path for each sequence, there are $\binom{2n}{n}$ such paths altogether. The number which don't leave the first quadrant are therefore, simply

$$\binom{2n}{n} - \text{(the number of paths which } do \text{ cross the } x\text{-axis).}$$

Thus, let us calculate this complementary number of paths which do wander into the 4th quadrant.

These complementary paths are precisely those which make contact with the line $y = -1$. (Such a path might bounce off $y = -1$, or cross it, or return to do either several times. Still, any path that makes contact with $y = -1$ certainly has left the first quadrant, and any path that leaves the first quadrant must extend at least as far as

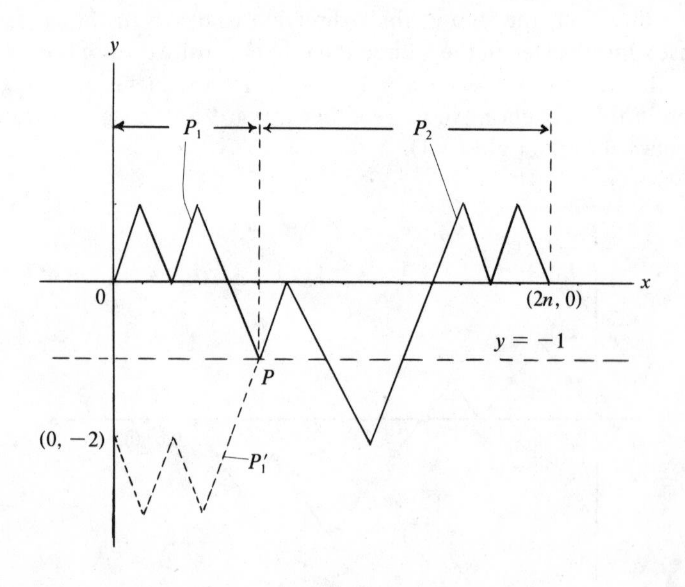

the line $y = -1$.) Let $P$ be the *first* point of contact of a path and the line $y = -1$ (that is, the first from 0), and suppose that the parts into which the path is divided by the point $P$ are $P_1$ and $P_2$. Now let $P_1$ be reflected in $y = -1$ to become a path $P_1'$ from the point $(0, -2)$ to the point $P$. Combining this image $P_1'$ with $P_2$ we obtain a path from the point $(0, -2)$ to the point $(2n, 0)$.

Conversely, let $Q$ be any such zigzag path from $(0, -2)$ to $(2n, 0)$. Since it starts and finishes on opposite sides of $y = -1$, it must cross this line somewhere. Let $T$ denote the *first* point of intersection with $y = -1$ from the endpoint $(0, -2)$, and let $T$ divide $Q$ into the parts $T_1$ and $T_2$. Clearly the image $T_1'$ of $T_1$ in $y = -1$ goes together with $T_2$ to yield a path from $(0, 0)$ to $(2n, 0)$, in particular a path between these points that crosses the $x$-axis. That is to say, there is a 1-1 correspondence between the subset of the paths from $(0, 0)$ to $(2n, 0)$ which leave the first quadrant and the entire set of all paths that join $(0, -2)$ and $(2n, 0)$. Since it is easy to count all the paths between two known points, we can now calculate the desired number of complementary paths.

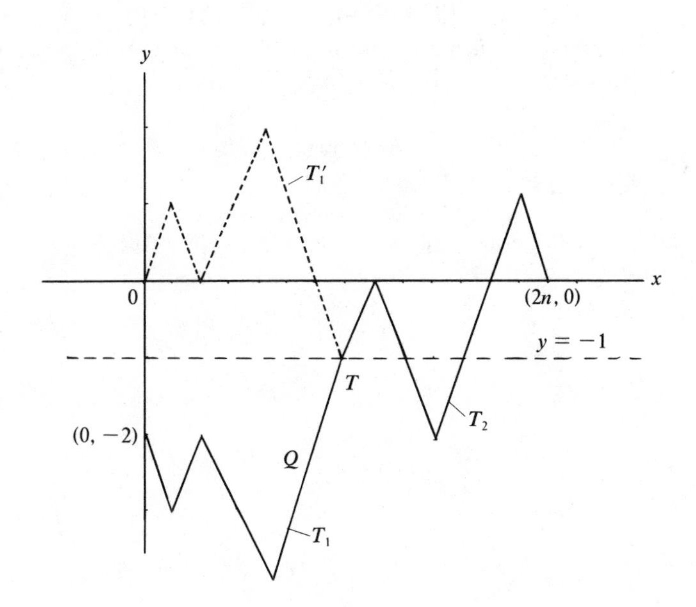

Since the "run" from $(0, -2)$ to $(2n, 0)$ is $2n$, every path from $(0, -2)$ to $(2n, 0)$ must possess $2n$ steps. In order to wind up 2 units above the starting level of the point $(0, -2)$, one must take 2 more upward steps than downward. This means that such a path must correspond to a sequence of $(n + 1)$ $+1$'s and $(n - 1)$ $-1$'s. The number of ways of ordering such sets of $+1$'s and $-1$'s is clearly

$$\binom{2n}{n-1},$$

the number of ways of picking $n - 1$ places for the $-1$'s.

Thus the sought-after number of paths that do not leave the first quadrant is

$$c_n = \binom{2n}{n} - \binom{2n}{n-1},$$

which simplifies in straightforward fashion to the conjectured

$$c_n = \frac{1}{n+1}\binom{2n}{n}.$$

The sequence $\{c_n\} = \{1, 2, 5, 14, 42, 132, 429, \ldots\}$ is called the Catalan numbers and is notorious for popping up all over the place in combinatorics.

### References

1. A. P. Street and W. D. Wallis, Combinatorial Theory: An Introduction, Charles Babbage Research Centre, Winnipeg, Manitoba, Canada, 1977.
2. D. Cohen, Basic Techniques of Combinatorial Theory, John Wiley & Sons, New York, 1978.

CHAPTER **10**

# FOUR CLEVER SCHEMES IN CRYPTOGRAPHY

I would like to tell you about four clever schemes in cryptography. In the first and third sections we will be dealing with the standard problem of foiling an eavesdropper, without regard to the possibilities of tampering or forgery. In the fourth section this larger problem will be challenged by a system that seems almost too good to be true. The second part concerns a procedure that might be used, for example, to maintain the anonymity of an electorate which, in certain cases, could have a profound bearing on one's future relationships.

## 1. Two Procedures Based on Discrete Logarithms

(i) *Passing a Key.* Since there are many easy ways to transform numbers, we assume that our message $m$ has already been translated into a positive integer $M$ by some system of numerical equivalents for the letters, numerals, and other symbols in the text. Anyone could invent a way to do this in a matter of seconds, and any straightforward system is satisfactory. We could have chosen to start disguising the message by using a complicated system in this initial stage, but we have not done so. Because a simple system would not long stand in the way of any competent eavesdropper, we acknowledge that knowing $M$ is as good as knowing $m$ itself.

Since our procedure calls for large-scale calculations that require electronic computers, it is convenient to express $M$ in the binary scale. While in practice $M$ could be thousands of digits long, it would be transmitted in blocks of a certain size, perhaps of length two or three hundred. The number of binary 300-tuples is $2^{300}$, about $10^{90}$, almost any of which can be selected to serve in the role of a "key." In order to illustrate how a key works, let us consider messages that are sent in blocks of 18 digits. Suppose the 64 6-digit binary numbers are used to encode the 52 lower-case and upper-case letters of the alphabet, the 10 numerals, and a couple of punctuation marks:

$$a = 000000, \qquad b = 000001, \ldots$$

The message "Yes," then, would translate into

$$M = \underbrace{110010}_{Y}\underbrace{000100}_{e}\underbrace{010010}_{s}.$$

Now any eavesdropper worth his salt would decipher this in an instant. Therefore, let us disguise our messages by reversing the digits (i.e., 0 to 1, or 1 to 0) in the positions occupied by the 1's in the 18-digit key $K$ below:

$$K = 011011101000101001.$$

This can be achieved simply by adding vectorially the binary 18-tuples $M$ and $K$ (which just means adding without carrying). The message that would be transmitted in this case would be the 18-tuple $T$:

$$\begin{array}{r} M = 110010000100010010 \\ K = 011011101000101001 \\ \hline T = 101001101100111011\,. \end{array}$$

In terms of our system of numerical equivalents, this represents "PS7." Since each symbol in $M$ has a 6-digit binary representation, there are 64 ways in which its digits may be altered by $K$ and, in general, the different sections of $K$ will treat differently any repeated occurrences of a symbol (for example, the key $K$ above changes YYY into PAB). Thus a simple key of this kind provides a great deal of protection for $M$. Because $K$ could be any of a multitude of $n$-tuples (except a few that are obviously "regular"), it is virtually impossible to discover the key by trial. And, at the other end, the recipient can recover the message $M$ from the received transmission $T$ merely by adding in the key once again (reversing the same positions restores the original). Thus, once a key has been communicated safely, it is very easy to carry on with relative security.

While it is virtually impossible to guess a key, clues to its makeup, at least parts of it, may be gained from scraps of information concerning the messages it is used to send. For example, the spy who begins every transmission with "Hello, this is Charlie." gives away the corresponding initial part of the key to anyone who becomes aware of his habit. In the absence of exact knowledge of a message, one can still systematically test educated guesses as to its content, thereby availing oneself of a real possibility, however improbable, of breaking the code. Thus it is advisable that the sender partition his blocks of coded text into subblocks and to convolute the subblocks according to some secret recipe. In this way an eavesdropper would not recognize the first subblock in the message. However, if the same key and the same scrambling of the subblocks were used over and over again, an eavesdropper might have a chance of putting two and two together. Thus things are made more difficult for an eavesdropper if these devices are changed fairly often; the least they can do is slow him down, which could be critical in certain cases.

The major reason, however, for changing the key and other encoding parameters is the very real danger of espionage. If this requires a confidential courier to visit a large number of places, it can be very expensive and is fraught with other dangers. In this first section we shall show how to pass a key virtually instantaneously, enabling one to adopt the wise precaution of changing the key with every message.

Of course, one might justify the expense of passing the first key on the basis that each succeeding key can be passed as a message using the preceding key. The weakness in this is that the discovery of any key by an outsider immediately reveals all later keys and would probably be a big help in determining the earlier ones. The discovery of a key which is communicated by the independent method we are about to propose would not jeopardize more than just its own messages.

The $2^n$ binary $n$-tuples can be thought of as the elements of the finite field $GF(2^n)$ [**1**]. Each element $(a_1, a_2, \ldots, a_n)$ in this field is identified with a polynomial of degree $n - 1$, $a_1 + a_2x + a_3x^2 + \cdots + a_nx^{n-1}$, whose coefficients are the components of the $n$-tuple. In $GF(2^3)$, for example, we have

$$000 = 0, \quad 100 = 1, \quad 010 = x, \quad 001 = x^2, \quad 110 = 1 + x,$$

$$101 = 1 + x^2, \quad 011 = x + x^2, \quad 111 = 1 + x + x^2.$$

While addition is given by the the usual component-by-component definition, multiplication is somewhat more involved. First, one chooses an irreducible polynomial $f(x)$ of degree $n$ having binary coefficients (a polynomial $g(x)$ is irreducible if it does not factor into the product of 2 or more polynomials of lesser degree whose coefficients are from the same field as the coefficients of $g(x)$; for the coefficients 0 and 1, the polynomial $x^n + x + 1$ is irreducible for many values of $n$ and is a favorite choice for $f(x)$; however, care must be taken, for it is not irreducible for every $n$). The product $uv$ of two $n$-tuples in $GF(2^n)$, then, is defined to be the $n$-tuple corresponding to the polynomial that is obtained by reducing, modulo $f(x)$, the product of the polynomials corresponding to $u$ and $v$. For example, suppose in $GF(2^3)$ that we take the modulus $f(x) = x^3 + x + 1$. With $x^3 + x + 1$ equivalent to zero, we have $x^3 = -x - 1 = x + 1$ (since $-1 \equiv +1 \pmod 2$). Repeatedly substituting $x + 1$ for $x^3$, a polynomial of any degree can be expressed as a polynomial of degree $\leq 2$, which identifies it with a 3-tuple of $GF(2^3)$. For example,

$$\begin{aligned} 011 \cdot 111 &= (x + x^2)(1 + x + x^2) \\ &= x + 2x^2 + 2x^3 + x^4 \qquad \text{(where } 2 \equiv 0 \pmod 2\text{)} \\ &= x + x^4 \\ &= x + x \cdot x^3 = x + x(x + 1) \\ &= x + x^2 + x = x^2; \end{aligned}$$

hence $011 \cdot 111 = 001$.

Now the $r - 1$ nonzero elements of $GF(r)$ always constitute a multiplicative group which is *cyclic* [**1**]:

$$g^0, g^1, g^2, \ldots, g^{r-2}.$$

Thus each element $g^s$ has an integral logarithm $s$ with respect to the generator $g$ as base (these are the "discrete" logarithms referred to in the title of this section). In terms of $g = 010 = x$ (which works for this example, but may not in general), we obtain the following representation for $GF(2^3)$:

$$\begin{aligned} 100 &= \quad 1 \quad &= x^0 \\ 010 &= \quad x \quad &= x^1 \\ 001 &= \quad x^2 \quad &= x^2 \end{aligned}$$

$$
\begin{aligned}
110 &= \quad 1 + x \quad &= x^3 \\
101 &= \quad 1 + x^2 \quad &= x^6 \\
011 &= \quad x + x^2 \quad &= x^4 \\
111 &= 1 + x + x^2 &= x^5;
\end{aligned}
$$

also

$$
\begin{aligned}
x^7 &= 1, \\
x^8 &= x, \\
x^9 &= x^2, \\
&\vdots \quad .
\end{aligned}
$$

While each nonzero element $u$ of $GF(2^n)$ assuredly has an integral logarithm $s$ ($u = g^s$), the general problem of calculating the logarithm of a given element $u$ is more than we can handle in 1983 if $n$ is of the order of a couple of hundred. Recent work by my colleagues Scott Vanstone, Ron Mullin, Ian Blake, and Ryo Fuji-Hara [4] has advanced the factoring of polynomials to the point where logarithms can be calculated quickly in $GF(2^{127})$. However, at present one could transmit binary 300-tuples without worrying that an eavesdropper would be able to calculate their discrete logarithms. As a result, $A$ and $B$ can safely pass a key, or any desired message $M$, in the form of a binary 300-tuple by adopting the following scheme, which needs no specification beyond a knowledge of its general procedures no matter how often it may be used, and which can even be revealed to an eavesdropper. An explanation of steps 3 and 4 follows the description.

1. $A$ generates a reasonably large random positive integer $a$, and sends to $B$ the 300-tuple $M^a$.
2. Similarly, $B$ generates a reasonably large random positive integer $b$, and returns to $A$ the 300-tuple $M^{ab}$. (If an eavesdropper were able to take logs, he would now be able to deduce $b$ from the relation $\log M^{ab} = b \log M^a$, and this would compromise the entire undertaking).
3. $A$ now removes his contribution to the confusion by raising $M^{ab}$ to the exponent $a^{-1}$ to yield $M^b$, which he sends to $B$.
4. Finally $B$ retrieves $M$ by raising $M^b$ to the exponent $b^{-1}$.

Steps 3 and 4 call for the extraction of the $a$th and $b$th roots of the received 300-tuples. Unfortunately, we must arrange for these operations to be possible. Consider the problem of extracting the 5th root of $x^2$ in $GF(2^3)$. We seek the element $x^k$ such that

$$(x^k)^5 = x^2.$$

Because $x^7 = 1$, this is equivalent to the congruence

$$5k \equiv 2 \pmod{7},$$

and we easily obtain $k = 6$.

Suppose, however, that we wish the cube root of $g^8$ in $GF(2^4)$. This time we seek $g^k$ such that

$$(g^k)^3 = g^8,$$

which requires

$$3k \equiv 8 \pmod{15} \qquad (g^{15} = 1 \text{ in } GF(2^4)),$$

i.e., integers $k$ and $q$ such that $3k - 8 = 15q$. Clearly there is no solution here because $3 \nmid 8$. In general, the $t$th root of $g^s$ exists in $GF(r)$ only if there is a solution to

$$(g^k)^t = g^s,$$

which is equivalent to $kt \equiv s \pmod{r - 1}$, $(g^{r-1} = 1$ in $GF(r))$, which has no solution if $s$ fails to share every common divisor of $t$ and $r - 1$.

As a result, we are encouraged to select a $GF(2^n)$ for which $2^n - 1$ is a prime number. This makes steps 3 and 4 possible for every choice of $a$ and $b$ (and these steps are practically trivial when they can be performed). The alternative is to eliminate from the possible values of $a$ and $b$ all the integers that are not relatively prime to $2^n - 1$. This is not that much of a bother, and still leaves $\varphi(2^n - 1)$ integers to pick from (where $\varphi$ denotes Euler's $\varphi$-function, [**2**]). Because $GF(2^{127})$ is no longer safe and the next $GF(2^n)$ for which $2^n - 1$ is a prime is $GF(2^{521})$, the latter alternative may be the more appealing for some time to come.

We conclude this first section with the following small-scale, but detailed, illustration. Today's computers can accomplish the entire 4-step procedure in a fraction of a second, even for quite lengthy $n$-tuples.

*A Detailed Example.* Suppose $A$ wants to send the message $M = 011$ to $B$.

1. $A$ picks a random number $a = 13$ and sends $M^a = 011^{13} = 110$ to $B$. ($A$ would not likely have gone to the trouble of selecting a generator $g$ and determined $M$ in the form $g^s$; he would probably compute $M^a$ by repeated squaring.)

$$M, M^2, M^4, M^8 \rightarrow M^{13} = M(M^4)(M^8) = 110.$$

(From our earlier work on $GF(2^3)$, we are in a position to verify this result as follows (recall $x^7 = 1$):

$$M = 011 = x^4, \qquad \text{giving} \qquad M^{13} = (x^4)^{13} = x^{52} = x^3 = 110.)$$

2. $B$ picks a random number $b = 11$ and sends $M^{ab} = 111$ to $A$. (Verification: $(x^3)^{11} = x^{33} = x^5 = 111$.)

3. Next, $A$ computes $a^{-1} = 13^{-1} = 6$ from $13k \equiv 1 \pmod 7$. Then $A$ determines $M^b$ from

$$M^b = (M^{ab})^{a^{-1}} = 111^6 = 001,$$

which he sends to $B$. (Verification: $(x^5)^6 = x^{30} = x^2$.)

4. Finally, $B$ calculates $b^{-1} = 11^{-1} = 2$ from $11k \equiv 1 \pmod 7$, and recovers $M$ from

$$M = (M^b)^{b^{-1}} = 001^2 = 011.$$

(Verification: $(x^2)^2 = x^4 = 011$.)

An eavesdropper would see only the messages 110 (from $A$ to $B$), 111 (back to $A$), and 001 (to $B$); unable to calculate the logarithms of these $n$-tuples (relative to any generator he might choose as base), he is unable to reconstruct the message $M = 011$, even though he knows that the transmitted messages are powers of the unknown $M$ (namely $M^a$, $M^{ab}$, and $M^b$), and, from the length of the blocks in the transmission, he knows the value of $n$.

(ii) *Preserving a Secret Ballot.* The problem of securing an election against cheating by voters, officials, and outsiders is a complicated and tricky business. In this second section we shall consider only one small aspect of this subject, an ingenious scheme, due to

Richard Lipton and Avi Wigderson (Princeton University), which permits an election to remain a secret ballot even though the votes are *sent in* to a central polling station [**3**].

If a vote can be cast in person, there is no difficulty in ensuring the secrecy of one's ballot. However, the communication of a vote involves the risk of interception and replacement or, at the very least, of jeopardizing the secrecy of your ballot. Modern systems of electronic communication certainly reduce the obvious risk of interception and replacement that must be accepted when using the mail or courier service. But, even with modern technology, in order to prevent one from voting more than once or usurping the votes of others, a voter is rightly required to identify himself before his vote is accepted. We will not consider how this might be done; let us assume that it has been accomplished to everyone's satisfaction. Having identified himself, it now seems to be impossible for our voter to communicate his preference without revealing to an election officer explicitly how he is voting. It seems futile to disguise his choice by using some kind of code, for eventually it must be decoded in order to be counted in the election, and it would probably be more trouble than it's worth to attempt to foil every means of tracing it back to its author. While it might appear to be impossible to disclose one's identity and not his vote, in fact it isn't! The way out is not to use a standard encoding technique to disguise one's vote but to use *other votes*. In the Lipton-Wigderson scheme each of the voters communicates to the central agency the results of the entire election; in communicating a vote, all the other votes serve as a disguise.

It might be difficult to see how this is a feasible approach, for it would appear that they are still up against the same old problem of preserving their anonymity in arranging for each of them to be able to disguise his vote with the votes of the others. All the details are given below, but the overall strategy here is similar to the procedure used in the first section, where deliberate, but organized, confusion is introduced to disguise the vital information of a message, to be removed later when doing so does not reveal any information that is not still disguised by the presence of at least other vital information. The end result is a summary of all the votes which contains no indication of who contributed what.

This scheme is not proposed for use with a large electorate. As in the first section, we assume that messages are communicated in the form of binary $k$-tuples. If $n$ is the number of voters and $r$ is the number of alternatives on the ballot, we require

$$nr < k.$$

Thus, using $GF(2^{300})$, we could handle an electorate of 60 or 70 voting on a slate of 4 alternatives. Again, the scheme hinges on the inability to calculate the discrete logarithm of an arbitrary element in the field $GF(2^k)$. To be on the safe side, then, let us take $k = 300$.

To set the notation, suppose that the $r$ alternatives on the ballot are each represented by a different polynomial of degree $\leq r$, having binary coefficients, and that the $n$ voters $A_1, A_2, \ldots, A_n$ express their preferences by selecting from these polynomials. There are always at least $r$ such irreducible polynomials **[1]**, and since we prefer to have a set of representatives in which no two share a common polynomial divisor, let us insist that each representative be irreducible (the reason for this will be made clear at the end). If the polynomial representing $A_i$'s vote is denoted by $x_i$, then the product of all $n$ of the selected polynomials,

$$X = x_1 x_2 \cdots x_n,$$

is of degree $\leq nr < k$, and hence can be represented by a single $k$-tuple having final components equal to 0 as far as necessary. Now then, if the voters will cooperate in carrying out the following 2-phase plan, they can successfully register their choices $x_i$ without forfeiting their anonymity. We will not consider the possible consequences of sabotage. It is remarkable enough that a system like this exists at all. It has a special character in that its price is only *cooperation*, not confidentiality; with no one ever knowing how anyone else voted, no one, including election officers, is required to keep any secrets other than his own.

*A Few Preliminaries.* We assume the electorate has been organized so that $A_i$ is able to receive $k$-tuples from $A_{i-1}$ and to forward $k$-tuples to $A_{i+1}$ ($A_n$ passes to $A_1$). If we imagine them gathered at their telecommunications terminals around the world, the election would be all over in a few minutes.

To begin with, each $A_i$ generates two large random positive integers $a_i$ and $b_i$ and selects a random binary $k$-tuple $y_i$; these choices are not to be disclosed to anyone else. The object of this exercise is to provide each voter with the $k$-tuple $X = x_1x_2 \cdots x_n$ containing the results of the entire election. Everybody gets this same result and in the process no one gets any information whose disguise he can penetrate.

$A_i$ begins to disguise his $x_i$ by multiplying it by his $y_i$. The basis of our procedure is a technique for determining the product of $n$ $k$-tuples, one from each of the $A_i$. Applying this process twice, once on the $x_iy_i$ and again on just the $y_i$, each voter is provided with the two products

$$x_1y_1x_2y_2 \cdots x_ny_n, \qquad y_1y_2 \cdots y_n,$$

from which $X = x_1x_2 \cdots x_n$ follows immediately by division. The process requires 3 steps, each involving $n$ stages.

*The Basic Process.*

(a) *Phase 1.* STEP 1. Step 1 calls for $A_i$ to receive a $k$-tuple from $A_{i-1}$, raise it to the power $a_i$, and pass it along to $A_{i+1}$. Beginning with his own $x_iy_i$ (which, of course, is not received from $A_{i-1}$), this operation is to be done $n$ times, thus circulating each $k$-tuple through the whole electorate (each of whom performs his operation on it), and returning it to its place of origin. For example, in the first pass

$A_i$ gives $(x_iy_i)^{a_i}$ to $A_{i+1}$;
next $A_{i+1}$ gives $(x_iy_i)^{a_ia_{i+1}}$ to $A_{i+2}$,
and so on...,
and at the conclusion of the round $A_i$ would receive
from $A_{i-1}$ the result $(x_iy_i)^{a_ia_{i+1}\cdots a_na_1a_2\cdots a_{i-1}}$.
If we let $a = a_1a_2 \cdots a_n$, then Step 1 provides
each $A_i$ with the $k$-tuple $(x_iy_i)^a$.

STEP 2. In this step, each person's task is simply to multiply what he receives by his $(x_iy_i)^a$ (which was just determined in Step 1), and pass it on, starting things off by passing on his $(x_iy_i)^a$ without alteration. Thus $A_i$ begins by passing $(x_iy_i)^a$ and eventually winds up with $(x_1y_1x_2y_2 \cdots x_ny_n)^a$.

STEP 3. Finally, each $A_i$ removes his contribution to the exponent by raising whatever he gets to the power $a_i^{-1}$ (this inverse is calculated from $a_i$ as in Section 1; thus $a_i$ and $b_i$ cannot be completely random, but must be relatively prime to $2^k - 1$). Accordingly, $A_i$ begins by passing

$$(x_1 y_1 \cdots x_n y_n)^{a a_i^{-1}}.$$

and ultimately obtains

$$x_1 y_1 x_2 y_2 \cdots x_n y_n.$$

(b) *Phase 2*. Applying the same 3-step procedure to the $y_i$ results in the product $y_1 y_2 \cdots y_n$. However, this time we must be sure to use the exponents $b_i$ in place of the $a_i$ (we will see why shortly).

As noted above, each $A_i$ can now calculate $X = x_1 x_2 \cdots x_n$, and it is this summary, containing his own vote as well as everybody else's, that is communicated to election headquarters.

The rest is routine, and can be carried out by each $A_i$ as well as the election officials. $X$ is a polynomial of degree $<k$ and is a product of $n$ factors chosen from the representative set of irreducible polynomials, which we shall denote by

$$P_1, P_2, P_3, \ldots, P_r.$$

It remains to determine how many of each $P_i$ are present in $X$. Accordingly, we attempt to divide $P_i$ into $X$. If it doesn't divide without remainder, we conclude that no one voted for alternative $i$. If it does divide, we try again to divide $P_i$ into the resulting $X/P_i$. The number of votes for alternative $i$ is simply the number of terms of the diminishing sequence $X$, $X/P_i$, $X/P_i^2$, $X/P_i^3$, ... that are divisible by $P_i$ (each time a factor $P_i$ is removed from $X$, the degree is reduced, making it easier for succeeding trials).

Because the $P_i$ are irreducible, $X$ must factor uniquely in the form

$$X = P_1^u P_2^v \cdots P_r^w.$$

Thus there is no chance of being misled into thinking that $P_i$ divides $X$ when in reality its factors are present only as parts of other $P_j$'s.

One nice feature of this system is that the possibility of official corruption is eliminated. The published results of the election must yield the common polynomial $X$ submitted by the electorate, or else treachery (or incompetence) is obvious to all.

Finally, we oberve that the $y$'s and $b$'s guard against a subtle but likely crucial possibility. If the $y$'s and $b$'s are neglected, then in Step 2 of Phase 1 $A_{i+1}$ would receive, in turn,

$$x_i^a,\ (x_{i-1}x_i)^a,\ (x_{i-2}x_{i-1}x_i)^a,\ \ldots,$$

from which he could compute each of the $x_j^a$; similarly, everyone could do this. Now, because the exponents here are all the same, one can infer that $x_i = x_j$ if $x_i^a$ is the same $k$-tuple as $x_j^a$, even though he can not compute $x_i$ directly. In this way each voter can compile a summary of which groups voted the same way. At the end, if it turns out that alternative $t$ got $s$ votes, then everybody would know who the $s$ people were who voted this way unless some other group of like-minded voters also happened to contain $s$ people. It is to foil calculations like this that the $y$'s were introduced; even if $x_i = x_j$, the chances are nil that $(x_i y_i)^a$ will be the same as $(x_j y_j)^a$.

Now, in determining the product of the $y$'s, which is necessary in order to get rid of them again, in Phase 2 each $A_i$ would be able to compute all the $y_i^a$ (as just described for the $x_i^a$). Combining with the similarly calculated $(x_i y_i)^a$ from Phase 1, he could then deduce all the $x_i^a$ and, in the final analysis, the $y$'s would not have helped at all. However, if in Phase 2 the $b$'s are used in place of the $a$'s, the difficulty vanishes, for the combination of $y_i^b$ and $(x_i y_i)^a$ is useless.

## 2. Public-Key Systems

In many cases, knowing how a message was encoded is as good as knowing how to decode it. However, some methods of encryption can be readily disguised so that the encoder remains unaware of how easy he is making things for the decoder. Our final two schemes concern public-key systems, in which a catalog of subscribers, revealing their personal methods of encryption, is distributed to all users (like a phone book).

I would like to begin by drawing your attention to two excellent expositions on these topics by researchers who know the subject intimately:

(a) Martin Hellman, The mathematics of public-key cryptography, Scientific American, August 1979;

(b) R. Rivest, A. Shamir, and L. Adleman, A method for obtaining digital signatures and public-key cryptosystems, MIT Memo MIT/LCS/TM-82.

These articles are not technically difficult and make very pleasant reading.

(i) *The Knapsack Problem.* The problem that faces an eavesdropper in this scheme is the innocent-sounding "knapsack problem," which simply asks one to find, from among a given set of numbers, a subset having a prescribed sum. For example, which of the integers

$$S = (1086, 708, 259, 589, 871, 1836, 82, 1747)$$

add up to 3756?

If the given set has only a few numbers, an exhaustive search is feasible. However, a set of 300 integers would have $2^{300}$ subsets, making such an attack impractical. The large-scale knapsack problem was, for some time, a major contender as the basis of a cryptographic system. However, the Israeli mathematician Adi Shamir showed in 1982 that it is not strong enough for this demanding role and its system has lost favor. Still, there is no reason why we can't enjoy the beautiful theoretical basis behind this proposal.

As usual, suppose the message $M$ is expressed in the binary scale. Now the encoding key that is listed in the public catalog for a subscriber $R$ consists of a $k$-tuple $E$ of positive integers:

$$E = (a_1, a_2, a_3, \ldots, a_k).$$

Any message that is sent to $R$ must be transmitted in blocks of length $k$. For example, suppose $E$ is the 8-tuple $S$ mentioned above and the first block of the message is

$$M = 0\,1\,1\,0\,1\,1\,1\,0.$$

Now $E$ and $M$ can be considered to be vectors of the same length, and as such they yield a certain scalar product $P$ (that is, add up the components in $E$ that occur in the places indicated by the 1's in $M$):

$$\begin{aligned} P = E \cdot M &= 0(1086) + 1(708) + 1(259) + 0(589) + 1(871) \\ &\quad + 1(1836) + 1(82) + 0(1747) \\ &= 3756. \end{aligned}$$

Since $M$ is a *binary* $k$-tuple, $P$ is a "knapsack sum" of the set $E$. It is this integer $P$ that is sent to $R$. Recovering $M$ from $P$ generally is a difficult knapsack problem for an outsider, but everything has been arranged to make it easy for $R$ to do so.

The system is based on two things: first, on the existence of a class of knapsack problems that, because of a special property, can be solved trivially; second, that instead of having to avoid such trivialities, there exist simple transformations that take us back and forth between them and knapsack problems that are genuinely difficult. Obviously the way to exploit this situation is to have the uncooperative side of these problems face the world and the congenial side face yourself.

The special class of easy problems consists of those for which each $a_i$ in the given set of numbers exceeds the sum of all the previous $a$'s:

$$\text{for all } i, \qquad a_i > a_1 + a_2 + \cdots + a_{i-1}.$$

This makes the recovery of $M$ a trivial exercise. For example, suppose the given set of numbers, which does possess the property in question, is $D = (6, 15, 31, 55, 112, 243, 529, 1015)$, and the given scalar product is 930. First we observe that the desired sum 930 lies between the numbers 529 and 1015 in $D$. Obviously the 1015 is too big to be a part of the sum. However, what about the 529? If is it *not* used, then the greatest available sum would be too small: even if all the earlier $a$'s are used, their sum is $<529$ which is $<930$. Therefore the 529 must be one of the addends, and it leaves a balance of 401. Similarly, unless the largest available number (243) is used toward this remainder, the undertaking is impossible to accomplish ($6 + 15 + 31 + 55 + 112 < 243 < 401$). Clearly, the greatest $a_i$ which does not exceed the current deficit must belong to the sum. Thus, moving from deficit to deficit, such a knapsack problem is solved with unexpected ease. In the example at hand, the process, pictured from right to left, would proceed as shown:

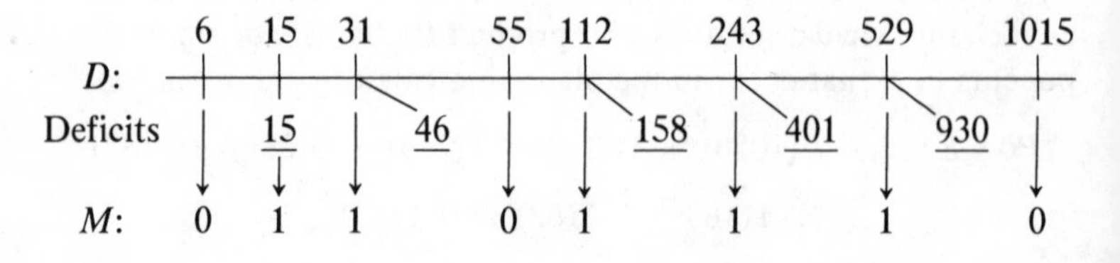

The examples that have been used can be put together to illustrate most of the knapsack system: The message $M = 01101110$ is encoded using $E = (1086, 708, 259, 589, 871, 1836, 82, 1747)$, transmitting to the recipient $R$ the scalar product $P = 3756$; $R$ transforms $P$ into $P' = 930$ (we will see shortly how this is done), and recovers $M$ from $P'$ by using his secret decoding set $D = (6, 15, 31, 55, 112, 243, 529, 1015)$. The question is, what are the transformations that accomplish this remarkable metamorphosis? The answer is—plain old modular arithmetic!

The encoding vector $E$ above is simply the special decoding vector $D$ transformed as follows: each $d_i$ in $D$ is multiplied by 850 and reduced, modulo 2007, to yield the corresponding $e_i$ in $E$; e.g., $d_1 = 6$ gives $6(850) = 5100 \equiv 1086 \pmod{2007}$, making $e_1 = 1086$. Such a transformation generally destroys the special property enjoyed by the numbers in $D$, with the result that $E$ inherits none of $D$'s remarkable amenity for the knapsack problem.

Almost any pair of relatively prime positive integers $(m, n)$ may be used instead of our multiplier $m = 850$ and modulus $n = 2007$. We insist only that $n$ exceed the sum of the integers in $D$.

If we denote the message and encoding vectors by

$$M = (m_1, m_2, \ldots, m_k),$$

and

$$E = (e_1, e_2, \ldots, e_k),$$

then the scalar product $P$ that is received by the decoder is

$$P = m_1e_1 + m_2e_2 + \cdots + m_ke_k.$$

Now suppose that it was the relatively prime positive integers $m$ and $n$ by which $E$ was derived from $D$ according to the stated relation

$$e_i \equiv md_i \pmod{n};$$

of course, $D = (d_1, d_2, \ldots, d_k)$ is chosen in the first place so that

$$d_i > d_1 + d_2 + \cdots + d_{i-1}, \qquad \text{for all } i.$$

In order to acquire the last key to the system, the decoder must calculate the inverse of $m \pmod{n}$, that is, $m^{-1}$ such that

$$mm^{-1} \equiv 1 \pmod{n}.$$

This is why $m$ and $n$ must be relatively prime. (This just requires a routine application of the Euclidean algorithm to the equation $m^{-1}m + qn = 1$, where $m$ and $n$ are known.) In the example at hand, it turns out that

$$m^{-1} = 961.$$

It is this inverse which is used to convert $P$ into the knapsack sum $P'$ to be mustered from the integers in $D$. In this transformation, the same modular procedure is used: $P$ is multiplied by $m^{-1}$ and the product is reduced (mod $n$):

$$P' \equiv Pm^{-1} \pmod{n}.$$

By "reduced" we mean to imply that $P' < n$. Noting that $e_i \equiv md_i \pmod{n}$, then, we have

$$\begin{aligned} P' &\equiv (m_1e_1 + m_2e_2 + \cdots + m_ke_k)m^{-1} \pmod{n} \\ &\equiv (m_1md_1 + m_2md_2 + \cdots + m_kmd_k)m^{-1} \pmod{n} \\ &\equiv m_1d_1 + m_2d_2 + \cdots + m_kd_k \pmod{n}. \end{aligned}$$

Since $n$ exceeds both sides of this congruence, this result is tantamount to an equality. Thus

$$P' = m_1d_1 + m_2d_2 + \cdots + m_kd_k,$$

and $M$ is found as the solution of a trivial knapsack problem.

To complete our example, we have

$$Pm^{-1} = 3756 \cdot 961 = 3609516 \equiv 930 \pmod{2007},$$

giving $M$ as the solution to the knapsack problem

$$6m_1 + 15m_2 + \cdots + 1015m_8 = 930.$$

(ii) *The RSA System.* Our final scheme is called the RSA system after its designers, the aforementioned Ronald Rivest, Adi Shamir, and Len Adleman.

(a) *How to Operate the System.* If you want to send a message to a subscriber $R$, the first thing you do is look up his personal encoding key in the public catalog. This consists of a pair of positive integers

$(e, n)$. The integer $e$ will be relatively small, but $n$ will be gigantic, perhaps a couple of hundred digits long. Because the base of the number system is not important, let us work in the ordinary decimal system. Now a message to $R$ must be sent in the form of a positive integer $<n$. Therefore, we would next attend to the routine task of translating the message into numerical code and dividing the result into blocks of appropriate size; suppose the first block is the integer $M$. $M$ is now encoded into its disguised form $M'$ by calculating the residue class, modulo $n$, of the power $M^e$:

$$M^e \equiv M' \pmod{n}.$$

(Of course, $M'$ is the representative of this class that lies in the range $[0, n-1]$.) For example, if the numbers involved were (the unreliably small)

$$(e, n) = (613, 274279) \quad \text{and} \quad M = 102114,$$

then the number that would be sent to $R$ is 245220 because

$$102114^{613} \equiv 245220 \pmod{274279}.$$

In order to recover $M$ from $M'$, an outsider would have to solve the congruence

$$x^{613} \equiv 245220 \pmod{274279}.$$

It is estimated that our fastest computers would take billions of years to solve the typical congruence $x^e \equiv M' \pmod{n}$ in the case of a 200-digit integer $n$.

However, at the other end, it is easy for $R$ to recover $M$ from $M'$ by using his secret decoding key $d$. Things have been arranged so that

$$(M')^d \equiv M \pmod{n}.$$

In the example, $d = 1333$ and $R$ would obtain $M$ from the calculation

$$245220^{1333} \equiv 102114 \pmod{274279}.$$

(b) *Signatures.* Of course, $R$'s decoding key $d$ only works on a message that has been encoded with his personal key $(e, n)$. Let us denote $R$'s encoding and decoding algorithms by $E_R$ and $D_R$ and express the results of their applications to $M$ and $M'$ by writing

$$E_R(M) = M', \qquad D_R(M') = M.$$

Obviously $D_R$ must be the inverse of $E_R$. However, the RSA system uses only pairs of keys in which each is the inverse of the other, that is,

$$\text{not only is} \quad D_R[E_R(M)] = M,$$

$$\text{but also} \qquad E_R[D_R(M)] = M.$$

While one can be fooled by a forged signature on a letter, a check, or a contract, the inverse character of $E$ and $D$ lies at the base of a foolproof means of identification. This ingenious procedure permits a banker to honor an "electronic check," or a businessman to accept a contract without fear of falsification. Not only can the recipient be sure that the message is genuine, but an objective third party, such as a magistrate, can also be totally convinced at any time in the future. In such transactions, of course, the sender needs protection, too. For example, he would not want to be robbed blind by a crooked banker who, in learning how to verify his signature on a check, learned at the same time how to forge it himself. In the system at hand, however, one need not worry about anything like this.

The marvellous idea behind it all is to have the sender $S$ not send $R$ the message $M$, but to send him the message $D_S(M)$; first he *de*codes $M$ using his own secret key $D_S$ and then, consulting the public catalog for $E_R$, sends $D_S(M)$ to $R$ in the usual way, namely, as

$$E_R[D_S(M)].$$

When $R$ gets this, he first applies his decoding key $D_R$ to obtain

$$D_R\{E_R[D_S(M)]\} = D_S(M).$$

Since $R$ is the only one in possession of $D_R$, he is the only one who can do this, ultimately verifying that the message is intended for him. Then $R$ recovers $M$ by applying $S$'s *en*coding key $E_S$, which he looks up in the catalog:

$$E_S[D_S(M)] = M.$$

Since the chance of $E_S$ decoding any message but one that has been encoded by $D_S$ is nil, $R$ can be quite sure that the message is from $S$.

Furthermore, $R$ can only claim the message contains material that is demonstrably obtainable from an application of $E_S$, as in the final step; thus any forged material must somehow be expressed in the form $D_S(M)$. Without knowing $D_S$, this is virtually impossible.

(c) *The Mathematical Basis of the System.* In the seventeenth century, Pierre de Fermat discovered that if a prime $p$ and a positive integer $a$ are relatively prime, then

$$a^{p-1} \equiv 1 \pmod{p}.$$

A century later, Euler found the more general relation

$$a^{\varphi(n)} \equiv 1 \pmod{n}$$

for relatively prime integers $a$ and $n$, where $\varphi(n)$ is his famous $\varphi$-function, which counts the number of positive integers $m \leq n$ that are relatively prime to $n$ [2]. If the prime decomposition of $n$ is

$$n = p_1^{a_1} \cdot p_2^{a_2} \cdots p_k^{a_k},$$

then a formula for $\varphi(n)$ is

$$\varphi(n) = n\left(1 - \frac{1}{p_1}\right)\left(1 - \frac{1}{p_2}\right) \cdots \left(1 - \frac{1}{p_k}\right).$$

In the case of $n = p$, a prime, we have

$$\varphi(n) = p\left(1 - \frac{1}{p}\right) = p - 1,$$

and if $n = pq$, the product of two different primes, we have

$$\varphi(n) = pq\left(1 - \frac{1}{p}\right)\left(1 - \frac{1}{q}\right) = (p-1)(q-1).$$

The first step in constructing one's encoding and decoding algorithms is to select two different primes $p$ and $q$, each of about 100 digits. There are zillions to choose from, and because the values of $p$ and $q$ must be kept secret, it is wise to select them by some random process. In fact, since they are so enormous, what else can we do but pick one of these giant integers at random and then see if it is a prime? With $p$ and $q$ in hand, $n$ is taken to be the product $pq$. This leads immediately to $\varphi(n) = (p - 1)(q - 1)$. Finally, $d$ and $e$ are taken to be a pair of multiplicative inverses, modulo $(p - 1)(q - 1)$:

$$de \equiv 1 \pmod{(p-1)(q-1)}, \quad \text{i.e.,} \quad de \equiv 1 \pmod{\varphi(n)}$$

(the Euclidean algorithm again). In this way, all the ingredients of a pair of keys $E(e, n)$ and $D(d, n)$ are obtained. The doubly-inverse nature that is required of our algorithms can be established from Fermat's theorem as follows.

Our desire is to show that

$$E[D(M)] = M = D[E(M)].$$

Since any value resulting from an application of $E$ and $D$ is in the range $[0, n - 1]$, as is $M$, the desired equalities follow immediately from the corresponding congruences

$$E[D(M)] \equiv M \equiv D[E(M)] \pmod{n}.$$

Now, by definition we have (mod $n$) that

$$E[D(M)] \equiv [D(M)]^e \equiv (M^d)^e \equiv M^{de}$$

and

$$D[E(M)] \equiv [E(M)]^d \equiv (M^e)^d \equiv M^{ed}.$$

It remains to show, then, that $M^{de} \equiv M \pmod{n}$.

For relatively prime $M$ and $p$, Fermat's theorem gives

$$M^{p-1} \equiv 1 \pmod{p}.$$

Since $\varphi(n) = (p - 1)(q - 1)$, then $M^{\varphi(n)} \equiv 1 \pmod{p}$, leading to $M^{k \cdot \varphi(n)+1} \equiv M \pmod{p}$, for any nonnegative integer $k$. But, clearly this holds for all $M$, even when $p|M$. Since $de \equiv 1 \pmod{\varphi(n)}$, it follows that, for some nonnegative integer $k$,

$$de = k \cdot \varphi(n) + 1.$$

Therefore, we have

$$M^{de} \equiv M \pmod{p} \quad \text{for all } M,$$

and $p \mid M^{de} - M$. Similarly, we have $q \mid M^{de} - M$, and because $p$ and $q$ are distinct primes, we obtain

$$pq \mid M^{de} - M,$$

which is the desired $M^{de} \equiv M \pmod{n}$.

(d) *The Problem of Security.* The security of a cryptographic system is a constant source of concern. The mathematical basis ensures that a system is workable but there is no guarantee that someone won't fathom its secrets tomorrow. One's vital secret in the present system is his value of $d$. Theoretically there is no difficulty in calculating $d$ from the public values of $e$ and $n$. The direct approach begins by rendering $n$ into its factors $p$ and $q$. From $p$ and $q$, $\varphi(n) = (p-1)(q-1)$ is determined and then the congruence $de \equiv 1 \pmod{\varphi(n)}$ is used. This requires each of $d$ and $e$ to be relatively prime to $\varphi(n)$. Consequently, the Euclidean algorithm yields $d$ as a solution $x$ to the equation

$$ex + \varphi(n) \cdot y = 1.$$

Once the factors $p$ and $q$ are known, the rest is routine. Our confidence in this system rests on the difficulty of factoring the large number $n$. By choosing $p$ and $q$ to have 100 digits or so, our present state of technology would take many years of computer time to factor their 200-digit product $n$. Breaking the system and factoring $n$ appear to be equivalent problems, but we can't even be sure of this. It would be nice if it were so, for in that case it might be feasible to obtain ongoing security simply by choosing bigger and bigger primes $p$ and $q$ so that factoring $n$ remained beyond our advancing technological capabilities.

At the present time, the RSA system sounds very promising, but the possibility that some unimaginably fast method of factoring will be discovered or some unnoticed property will be recognized as a fatal flaw are very real dangers. For example, even without a general method for finding $d$, an outsider might be able to turn up an individual message $M$ simply by checking the values of $E(M)^r \pmod{n}$ as $r$ runs through a relatively small range of values $(1, 2, \ldots)$. While $r = d$ will certainly decode $E(M)$, the congruence $E(M)^r \equiv M$ (mod

$n$) might also have solutions in which $r$ is much smaller than $d$. This depends partly on the message $M$ as well as $n$. To illustrate with a very small example, suppose that

$$n = pq = 3 \cdot 11 = 33, \qquad \varphi(n) = 2 \cdot 10 = 20,$$

$$e = 3, \qquad d = 7 \qquad (de \equiv 1 \pmod{20}).$$

Encoding the message $M = 09$, we determine $M^e = (09)^3 = 729 \equiv 3 \pmod{33}$. Then the values of $x$ produced by $E(M)^r \equiv 3^r \equiv x \pmod{33}$ as $r = 1, 2, 3, \ldots$ are $x = 3, 9, 27, \ldots$, which turns up the message $M = 09$ at the second trial (while $d = 7$). In fact the message $M = 12$ encodes into $E(M) = 12$, implying that the transmitted $E(M)$ is simply $M$ itself.

One might wonder how an outsider would recognize the second value $x = 09$ as the original message $M$. Of course, it might escape him, but he can try translating every $x$ according to various standard tables of numerical equivalents in the hope of getting a message that makes sense. Even if the encoded message is an especially secure one of the type $D_S(M)$, he is forced only to an extra step of treating each $x$ with the public algorithm $E_S$.

Obviously such weaknesses must be eliminated in order to make a system practical. This is not an easy area, but the RSA system seems to be strengthened by choosing

(i) $d$ to be a large prime,
(ii) $p$ and $q$ so that each of $p - 1$ and $q - 1$ has a large prime factor,
(iii) the greatest common divisor of $p - 1$ and $q - 1$ is relatively small.

It is also advisable to choose $p$, $q$, and $d$ by some random process in order to help guard against their discovery by a direct approach. In the realm of 100-digit integers, there are so many primes that this is entirely feasible. The problem of discovering a 100-digit prime is not negligible, but it is not at all of the same order of difficulty as that of factoring a 100-digit integer.

(e) *Two Notes.* The small example given in part (a) of this section was constructed from the primes

$$p = 157, \qquad q = 1747,$$

giving

$$n = 274279, \qquad \varphi(n) = 272376,$$

and allowing the choice of

$$d = 1333, \qquad e = 613.$$

Finally, it is a pleasure to extend sincere thanks to my friend and colleague Scott Vanstone for introducing me to this subject and for encouraging and assisting me throughout the writing of this essay.

## References

1. R. Lidl, H. Niederreiter, Finite Fields, Vol. 20, Encyclopedia of Mathematics and Its Applications, Addison-Wesley, Reading, Mass., 1983.
2. W. Sierpiński, Theory of Numbers, Vol. 42, Monographs in Mathematics, Państwowe Wydawnictwo Naukowe, Warsaw, 1964.
3. R. J. Lipton, A. Wigderson, Multi-Party Cryptographic Protocols (extended abstract), to appear.
4. I. F. Blake, R. Fuji-Hara, R. C. Mullin, S. A. Vanstone, Computing logarithms in finite fields of characteristic 2, Siam Journal of Algebraic and Discrete Methods, 5(2) (June 1984).
5. W. Diffie, M. Hellman, New directions in cryptography, IEEE Transactions on Information Theory (Nov. 1976) 644–654.
6. M. Gardner, Mathematical Games, Scientific American, 236 (February 1977) 120–124.
7. G. J. Simmons, Cryptology, the mathematics of secure communications, The Mathematical Intelligencer, 1(4) (1979) 233–246.

CHAPTER **11**

# GLEANINGS FROM NUMBER THEORY

In this section let us take a leisurely look at a few brief items from number theory.

1. Here's a little problem that might keep you going for a minute or two.

> For each positive integer $n$, determine a set of $n$ distinct positive integers having the property that no subset of them adds up to a perfect square. (*Crux Mathematicorum*, 1976, p. 29)

Since no perfect square ends in an odd number of 0's, any $n$ different odd powers of 10 will do.

The same idea provides a lovely way of completing the popular argument for the irrationality of $\sqrt{2}$ from the point $2a^2 = b^2$; in base 2, the left side ends in an odd number of 0's while the right side ends in an even number, giving the desired contradiction.

2. Isn't it interesting that, for all positive integers $n$,

$$1 - \frac{1}{2} + \frac{1}{3} - \frac{1}{4} + \frac{1}{5} - + \cdots + \frac{1}{2n-1} - \frac{1}{2n}$$

$$= \frac{1}{n+1} + \frac{1}{n+2} + \cdots + \frac{1}{2n}?$$

The following simple but very perceptive derivation by Judith Lum Wan (James Cook University of North Queensland, Australia) was reported in *Crux Mathematicorum*, 1980, p. 282.

Clearly the left side is given by

$$1 + \frac{1}{2} + \frac{1}{3} + \frac{1}{4} + \cdots + \frac{1}{2n}$$

$$- 2\left(\frac{1}{2} + \frac{1}{4} + \frac{1}{6} + \cdots + \frac{1}{2n}\right)$$

$$= 1 + \frac{1}{2} + \cdots + \frac{1}{2n} - \left(1 + \frac{1}{2} + \frac{1}{3} + \cdots + \frac{1}{n}\right)$$

$$= \frac{1}{n+1} + \frac{1}{n+2} + \cdots + \frac{1}{2n}.$$

3. Everybody is familiar with those puzzles that call for a specified positive integer to be expressed in terms of a prescribed set of symbols, such as four 4's. A popular version of this kind of puzzle requires the use of each digit 0, 1, . . . , 9 exactly once. What a marvelous stroke it was by Verner Hoggatt Jr. to settle the cases of all positive integers $n$ at one blow:

$$\log_{\frac{0+1+2+3+4}{5}} [\log_{\underbrace{\sqrt{\sqrt{\cdots\sqrt{-6+7+8}}}}_{n \text{ roots}}} 9] = n.$$

On top of that, when this expression is written in the usual way, I'll be darned if the digits aren't put down in their natural order 0, 1, . . . , 9!

In more compact form this equation asserts that

$$\log_2[\log_{9^{1/2^n}} 9] = n,$$

which is equivalent to

$$\log_{9^{1/2^n}} 9 = 2^n,$$

and finally

$$(9^{1/2^n})^{2^n} = 9$$

(*Crux Mathematicorum*, 1980, p. 284).

4. Let $C$ denote the set of positive integers which, when written in base 3, do not contain the digit 2. Prove that no 3

members of $C$ can be in arithmetic progression. (*Spectrum*, published by the University of Sheffield; solution by S. R. Blake of Rugby School)

Setting aside the numbers $C$, consider any 3 positive integers in arithmetic progression, $a$, $a + d$, $a + 2d$, and suppose they are expressed in base 3. Since $d$ is not zero (the numbers are different), it must contain some nonzero digit (i.e., 1 or 2). Let the last nonzero digit of $d$ occur in the $n$th place. Then, regardless of the $n$th digit of $a$, the $n$th digits of the 3 numbers $a$, $a + d$, $a + 2d$ must be the 3 different digits 0, 1, 2 (since all later digits in $d$ are 0's, when adding $d$ there is no carry over to the $n$th position from the previous addition).

Table of $n$th Digits

| $d$ | $a$ | $a + d$ | $a + 2d$ |
|---|---|---|---|
| 1 | 0 | 1 | 2 |
| 1 | 1 | 2 | 0 |
| 1 | 2 | 0 | 1 |
| 2 | 0 | 2 | 1 |
| 2 | 1 | 0 | 2 |
| 2 | 2 | 1 | 0 |

Thus at least one of $a$, $a + d$, or $a + 2d$ must contain a digit 2 and fail to belong to the set $C$, implying the desired conclusion.

5. Partition the positive integers into two subsets so that each contains arithmetic progressions of every finite length but neither contains an arithmetic progression of infinite length. (*Spectrum*)

First partition the ordered sequence of positive integers into subsets in which the number of integers are, respectively, 1, 2, 3, . . . :

$$(1), (2, 3), (4, 5, 6), (7, 8, 9, 10), (11, 12, 13, 14, 15), \ldots.$$

Then construct the desired subsets A and B by assigning these groups alternately to them:

A: (1; 4, 5, 6; 11, 12, 13, 14, 15; ...)

B: (2, 3; 7, 8, 9, 10; 16, 17, 18, 19, 20, 21; ...).

In this way each of $A$, $B$ contains ever-increasing intervals of consecutive integers, implying consecutive progressions of arbitrary finite length; also, each of $A$, $B$ is perforated with ever-increasing gaps of consecutive integers, showing that no common difference $d$ is great enough to extend an arithmetic progression indefinitely far along the subset (eventually a gap exceeding $d$ must be encountered, and all later gaps are even bigger).

6. The great French mathematician Joseph Lagrange proved that every positive integer is a sum of 4 or fewer perfect squares. Obviously not every integer needs as many as 4 squares, and sometimes it turns out that expressing an integer as a sum of more than 4 squares can be a problem. For example, 14 is the sum of the 5 squares $4 + 4 + 4 + 1 + 1$, but there is no way to do the same for 15. I am not sure what the greatest integer is that defies expression as a sum of 5 nonzero squares (I think it's 33), but there is a neat way of showing that it cannot exceed 169 (due to Ivan Niven and Herbert Zuckerman).

If $n > 169$, then we have, by Lagrange's theorem, that

$$n - 169 = a^2 + b^2 + c^2 + d^2,$$

where not more than 3 of $a$, $b$, $c$, $d$ are zero. This yields

$$\begin{aligned} n &= a^2 + b^2 + c^2 + d^2 + 13^2 && \text{(if none is zero)} \\ &= a^2 + b^2 + c^2 + 12^2 + 5^2 && \text{(if one is zero)} \\ &= a^2 + b^2 + 12^2 + 4^2 + 3^2 && \text{(if two are zero)} \\ &= a^2 + 10^2 + 8^2 + 2^2 + 1^2 && \text{(if three are zero).} \end{aligned}$$

7. Now let's consider the second problem from the 1982 British Olympiad.

> If a multiple of 17, when expressed in base 2, contains exactly 3 1's, prove that it must have at least 6 0's and

that if it were to have exactly 7 0's, then it would have to end in a 0.

(a) If the binary representation of 17k contains exactly 3 1's, then

$$17k = 2^a + 2^b + 2^c \quad \text{for some } a > b > c \geq 0.$$

Then

$$17k = 2^c(2^{a-c} + 2^{b-c} + 1),$$

implying that

$$2^c | k.$$

If $k = 2^c m$, then

$$17m = 2^{a-c} + 2^{b-c} + 1,$$

and we see that $m$ must be odd.

Now if 17k contains 9 or more digits altogether, that is, if $a \geq 8$, then at least 6 of the digits must be 0's (since 3 digits are 1's). Proceeding indirectly, suppose $a \leq 7$. In this case, $a - c \leq 7$, and the smaller $b - c$ must be $\leq 6$, giving

$$17m \leq 2^7 + 2^6 + 1 = 193.$$

Thus $m \leq 193/17$, making $m \leq 11$, and we have that $m = 1, 3, 5, 7, 9$, or 11. But we shall see that none of these values is feasible, implying the desired conclusion.

The equation

$$17m - 1 = 2^{a-c} + 2^{b-c}$$

shows that $17m - 1$ is the sum of two different numbers from the list (2, 4, 8, 16, 32, 64, 128). But, as $m$ varies, $17m - 1$ takes the values 16, 50, 84, 118, 152, and 186, none of which is the sum of two of (2, 4, 8, 16, 32, 64, 128).

(b) If the representation were to possess exactly 7 0's, then 17k would have to be a 10-digit number consisting of 7 0's and 3 1's. Then "$a$" would be 9 and

$$17m = 2^{9-c} + 2^{b-c} + 1.$$

Now if $c = 0$, then

$$17m = 2^9 + 2^b + 1, \qquad \text{where } 0 = c < b \leq 8.$$

The variations in $b$ place $17m$ among the 8 numbers

$$(515, 517, 521, 529, 545, 577, 641, 769).$$

But none of these is divisible by 17. Therefore $c > 0$, and we conclude that $17k$ must end in a 0.

8. The following was posed by Norman Swanson (South Hamilton, Massachusetts) as Problem 204 in the *Two-Year College Mathematics Journal*, 1982, p. 337.

> Determine all pairs of positive integers $(m, n)$ such that $(n - 1)!n! = m!$.

Dividing by $n!$, we get

$$(n - 1)! = (n + 1)(n + 2) \cdots m.$$

Now if any factor $n + i$ on the right side were to be a prime number, it could not be matched on the left side as a product taken from the smaller numbers $1, 2, \ldots, n - 1$. Thus all the factors $n + i$ on the right side must be composite numbers.

Suppose that $p$ is the greatest prime number $\leq n - 1$, that is, the greatest prime factor on the left side. Thus in our equation, all the factors beyond the factor $p$ on the left must be composite numbers. However, because $p$ divides the left side, it must also divide the right side, implying that some factor on the right must be of the form $rp$, where $r \geq 2$.

$$1 \cdot 2 \cdot 3 \cdots p(p + 1) \cdots (n - 1) = (n + 1)(n + 2) \cdots (rp) \cdots m$$

| ···················· only composite factors ···|

Now, by the well-established theorem known as Bertrand's Postulate, there exists a prime number between $p$ and $2p$. Therefore, at least one of the integers between $p$ and $rp$ must be a prime. The only one we have not already seen as composite is the number $n$, itself, which is absent from our equation. Thus we have the little corollary that $n$ must be a prime. It is clear, however, that $n$ is the *only* prime between $p$ and $2p$.

This forces $p$ to be less than 6, for a stronger version of Bertrand's Postulate declares that, for $k \geq 6$, there exist at least two prime numbers between $k$ and $2k$. Therefore the only possibilities for $p$ and $n$ are (2, 3), (3, 5), and (5, 7), and by trial the only feasible case here is (5, 7), yielding the solution $(m, n) = (10, 7)$:

$$6!7! = 10!.$$

Of course, for $n = 1$ or 2, there is no prime number $\leq n - 1$. Considering these cases separately, we also obtain the trivial solutions (1, 1) and (2, 2).

(This solution is the result of a collaboration with Bruce Richter, University of Waterloo.)

9. The following solution to Problem E2752 in the *American Mathematical Monthly* (1979, p. 56) and problem (b) below (with solution) were related to me by Herb Shank (University of Waterloo).

(a) Problem E2752. ($[x]$ denotes the greatest integer $\leq x$.) If $a, b, c, d$ are positive real numbers such that

$$[na] + [nb] = [nc] + [nd]$$

for all positive integers $n$, prove that

$$a + b = c + d.$$

Let $\{x\}$ denote the fractional part of $x$:

$$0 \leq \{x\} < 1 \quad \text{and} \quad x = [x] + \{x\}.$$

Then

$$[na] = na - \{na\}$$

and

$$\frac{[na]}{n} = a - \frac{\{na\}}{n},$$

and so on.

Then, dividing the given relation by $n$, we obtain

$$a - \frac{\{na\}}{n} + b - \frac{\{nb\}}{n} = c - \frac{\{nc\}}{n} + d - \frac{\{nd\}}{n}$$

$$(a + b) - (c + d) = \frac{1}{n}(\{na\} + \{nb\} - \{nc\} - \{nd\}).$$

In view of $0 \leq \{x\} < 1$, it follows that

$$|(a + b) - (c + d)| < \frac{2}{n}.$$

Since this holds *for all* $n$, it must be that

$$(a + b) - (c + d) = 0,$$

as required.

(b) We conclude with a related problem that is reminiscent of Beatty's Theorem [**1**].

If $a, b, c, d$ are positive *irrational* numbers such that

$$a + b = c + d,$$

prove that $[na] + [nb] = [nc] + [nd]$
for all $n = 1, 2, 3, \ldots$.

Let $a + b = c + d = k$. Clearly

$$\begin{aligned}[na] + [nb] &= na - \{na\} + nb - \{nb\} \\ &= n(a + b) - (\{na\} + \{nb\}) \\ &= nk - (\{na\} + \{nb\}),\end{aligned}$$

showing that $(\{na\} + \{nb\})$ is an *integer*. Because $a$ and $b$ are irrational, $na$ and $nb$ are never integers, implying that their fractional parts are never 0, and we have

$$0 < \{na\} + \{nb\} < 2.$$

Thus it must be that $\{na\} + \{nb\} = 1$, and $[na] + [nb] = nk - 1$. Similarly,

$$[nc] + [nd] = nk - 1,$$

and the desired conclusion follows.

### References

1. Ross Honsberger, Ingenuity in Mathematics, New Mathematical Library Series, Mathematical Association of America 1970; see p. 93 for Beatty's Theorem.

CHAPTER **12**

# SCHUR'S THEOREM: AN APPLICATION OF RAMSEY'S THEOREM

As we saw in an earlier chapter, a graph on $n$ vertices which contains all possible $\binom{n}{2}$ edges is called a complete graph and is denoted by $K_n$.

Suppose that each of the 15 edges of a complete graph $K_6$ is colored either red or blue. It is well known that, no matter which edges are colored red and which ones blue, a triangle must be produced that has all three sides the same color. However, this is not true for every way of coloring the edges of a $K_5$. Similarly, if the edges of a $K_{17}$ are each colored one of three colors, a "monochromatic" triangle always results, but this is not true for every 3-coloring of the edges of a $K_{16}$. More generally, for a set of $n$ colors, there exists a cutoff number $r(n)$ such that every $n$-coloring of the edges of a $K_{r(n)}$ must contain a monochromatic triangle, while this is not the case for every $n$-coloring of a $K_{r(n)-1}$. The existence of these minimum numbers $r(n)$ was established by the British mathematician F. P. Ramsey in 1930. In fact, his theorem is a far-reaching generalization of these simple considerations (in which we have discussed only monochromatic *triangles* obtained by coloring the *edges* of a $K_n$). The numbers $r(n)$, which are the subject of the present discussion, comprise only a small fraction of the entire class of generalized Ramsey numbers.

It should be noted that, although Ramsey numbers have long been known to exist, very little is known about their values. For example, in addition to $r(2) = 6$ and $r(3) = 17$ (for 2 and 3 colors), it is known only that $r(4)$ is at least 42 and not greater than 66 (see [**2**]). Fortunately, the shortage of numerical data has not prevented the discovery of many interesting properties of Ramsey numbers.

**Schur's Theorem**

If the numbers (1, 2, 3, ..., 13) are divided into the three subsets

$$(1, 4, 10, 13), \quad (2, 3, 11, 12), \quad (5, 6, 7, 8, 9),$$

you will find that no two numbers (the same or different) in the same subset add up to another number in that subset. However, there is no escaping this result no matter how the first 14 (or more) positive integers are divided into three subsets.

If one divides the numbers $(1, 2, 3, \ldots, k)$ into just two subsets, then the property in question sets in at $k = 5$. For example, (1, 2, 3, 4) can be divided into (1, 4) and (2, 3), but (1, 2, 3, 4, 5) cannot be partitioned successfully. In general, there exists a minimum number $s_n$ at which the property goes into effect, that is, no matter how one partitions the numbers $(1, 2, 3, \ldots, s_n)$ into $n$ subsets, one of the subsets will contain three numbers $x, y, z$ ($x$ and $y$ not necessarily different) such that

$$x + y = z.$$

With the trivial $s_1 = 2$, we have $s_2 = 5$ and $s_3 = 14$.

No general formula for $s_n$ is known. The Ramsey numbers $r(2) = 6$ and $r(3) = 17$ are close but certainly not equal to $s_2$ and $s_3$. However, in 1916, I. Schur established the general relation

$$s_n \leq r(n) \quad \text{for all } n = 1, 2, 3, \ldots.$$

*Proof.* Suppose that the numbers $(1, 2, 3, \ldots, r(n))$ are partitioned into $n$ subsets $(S_1, S_2, \ldots, S_n)$ in any way whatever.

Let the vertices of a $K_{r(n)}$ be numbered $1, 2, \ldots, r(n)$, and let each edge be labelled with the difference $|i - j|$ between the numbers $i$ and $j$ that occur at its end-vertices. Thus the edge joining vertices 4 and 9 is labelled 5, as is the edge joining vertices 7 and 12, and so on. Also, let a set of $n$ colors be numbered $1, 2, \ldots, n$. Finally, let every edge that bears a label belonging to the subset $S_i$ be colored with color $i$ (if the number 8 belongs to $S_5$, then every edge bearing an 8 is done with color 5). In this way each edge of the $K_{r(n)}$ is colored with one of the $n$ colors.

Now for every triangle $ijk$ in our configuration, it is easy to see that the sum of the labels on two of the sides must be the label on the third side. Because $i, j, k$ label vertices, they are different numbers, and must have some ordering by magnitude, say $i > j > k$. Then the labels on the sides of $\Delta ijk$ are

$$i - j, j - k, i - k \quad \text{(which may not all be distinct)},$$

and

$$(i - j) + (j - k) = i - k.$$

But the labels on the edges are all members of the subsets $S_1, S_2, \ldots, S_n$. Thus every triangle provides us with three numbers which have the addition property in question. And the three numbers will all belong to the same subset $S_i$ if and only if the three sides of the triangle all have the same color ($i$). Since some triangle must be monochromatic, according to Ramsey's theorem, the proof is complete.

**References**

1. J. A. Bondy and U. S. R. Murty, Graph Theory With Applications, American Elsevier, New York, 1976, p. 112.
2. H. L. Dorwart and D. Finkbeiner, Chromatic Graphs, in Mathematical Plums, Dolciani Mathematical Expositions, Vol. 4, Mathematical Association of America, 1979.

CHAPTER **13**

# TWO APPLICATIONS OF HELLY'S THEOREM

The *diameter* of a set of points is the least upper bound of the distances determined by pairs of the points of the set. Thus the diameter of the points in a unit square, for example, is $\sqrt{2}$ (whether or not the set contains its boundary). The diameter is not defined in terms of the maximum distance because the set might not give rise to a maximum distance (e.g., the interior of a square), whereas the least upper bound always exists.

In our first application, we shall establish a remarkable property of the sets of points in the plane which have diameter $\leq 1$. This is due to H. W. E. Jung (1901).

JUNG'S THEOREM. *Any plane set of diameter $\leq 1$ can be enclosed in a circle of radius $1/\sqrt{3}$.*

The set of points determined by an equilateral triangle of side 1 is a set of diameter 1, and it is easy to calculate that its circumradius is exactly $1/\sqrt{3}$. Thus we note that the covering circle in Jung's theorem must contain its circumference and that it is the smallest *circle* which is capable of covering all plane sets of diameter $\leq 1$.

A nice proof of Jung's theorem can be based on a very versatile theorem which was published by Edward Helly in 1923. It concerns collections of convex sets. A set $S$ is said to be convex if, for every pair of its points $A$ and $B$, the entire segment $AB$ belongs to $S$. Helly's theorem makes the following claim.

HELLY'S THEOREM. *If, for each choice of three sets from a collection of $n$ convex sets in the plane, the three sets have a point in common, then there is a point which is common to all $n$ of the sets.*

This theorem is true for all finite $n$, but there are infinite collections for which it fails. However, it is valid for an infinite collection if even one of the sets is bounded. It is to such infinite collections that we shall apply the theorem.

The proof of Helly's theorem is not particularly long and is certainly not difficult. However, it is perhaps not as exciting as its many applications, and we shall not consider it here. Complete details for both the finite and infinite cases can be found in these truly outstanding books:

1. Yaglom and Boltyansky, Convex Figures, Holt, Rinehart, and Winston (the finite case).
2. Hadwiger, Debrunner, and Klee, Combinatorial Geometry in the Plane, Holt, Rinehart, and Winston (the infinite case).

Let us use Helly's theorem to prove Jung's engaging result.

*Proof of Jung's theorem.* Let $S$ denote any set of points in the plane having diameter $\leq 1$. About each point of $S$ as center, construct a circle of radius $1/\sqrt{3}$. The collection of circles thus obtained comprises a finite or infinite collection of convex sets to which Helly's theorem can be applied once we have shown that any three circles have a common point. Consider, therefore, any three circles of the collection, say those having centers $A$, $B$, $C$ (Figure 1).

Since the sum of the angles of a triangle is $180°$, not all three of the angles can be $< 60°$. In triangle $ABC$, suppose angle $C \geq 60°$. By

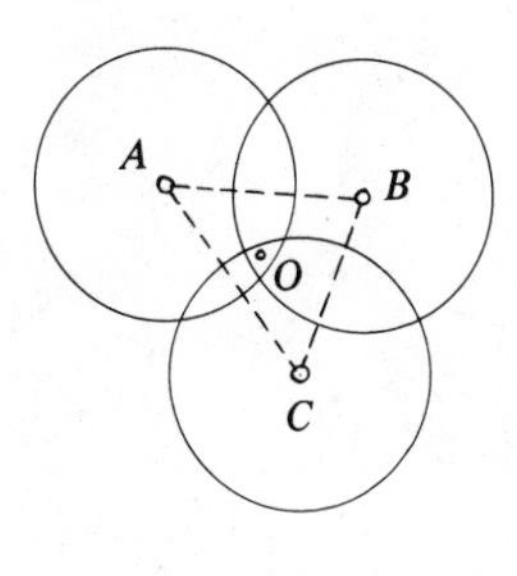

Figure 1

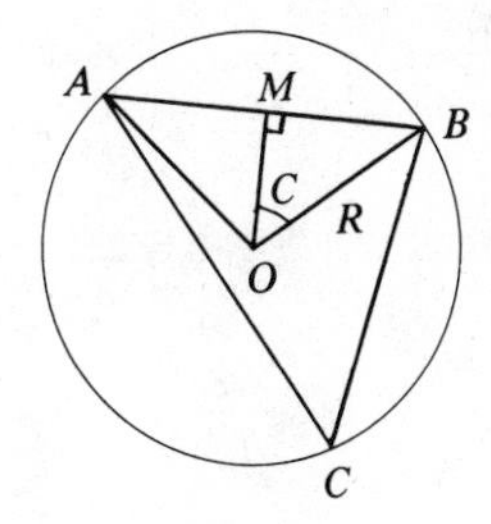

Figure 2

the well-known trigonometric formula for the circumradius $R$ of a triangle, we obtain $R = AB/2 \sin C$ (Figure 2).

Because $S$ is a set of diameter $\leq 1$, we have $AB \leq 1$, and we get

$$R \leq \frac{1}{2 \sin C}.$$

Now angle $C$ is either $> 120°$ or not. In the event that $60° \leq C \leq 120°$, we have

$$\sin C \geq \sin 60° = \frac{\sqrt{3}}{2},$$

and it follows that

$$R \leq \frac{1}{2\left(\frac{\sqrt{3}}{2}\right)} = \frac{1}{\sqrt{3}}.$$

That is to say, the circumcenter $O$ of triangle $ABC$ is not farther from each center $A$, $B$, $C$ than the radius $1/\sqrt{3}$ of the circles drawn about these points. Therefore the circumcenter $O$ is a point that is common to these three circles.

In the case of $C > 120°$, it is clear (Figure 3) that $C$ must lie inside the circle having $AB$ as diameter (this is true for all $C > 90°$). If $O$ denotes the midpoint of $AB$, then we have

$$\text{the radius} = \frac{1}{2} AB \leq \frac{1}{2},$$

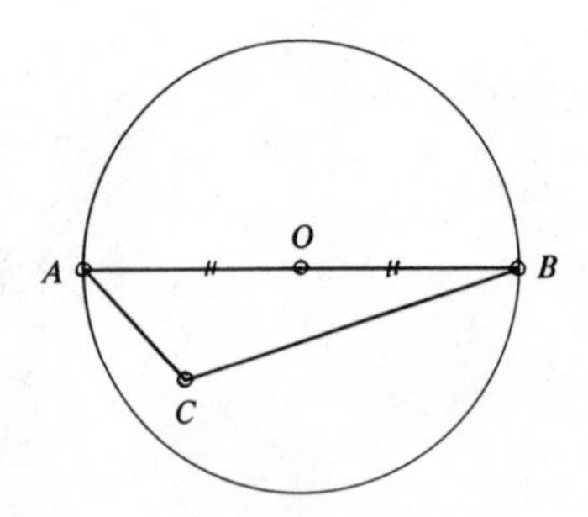


Figure 3

and

$$OC < \frac{1}{2}.$$

Because $1/2 < 1/\sqrt{3}$, the point $O$ must be closer to $A$, $B$, $C$ than $1/\sqrt{3}$, implying again that $O$ is common to the three circles around $A$, $B$, $C$. Clearly, the case when $A$, $B$, $C$ are collinear is also covered by this argument.

In all cases, then, the three circles have a point in common and, by Helly's theorem, there exists some point $X$ which lies in all the circles of the collection ($X$ may not belong to $S$). It is easy to see that the circle of radius $1/\sqrt{3}$ having center $X$ must cover all the points of $S$. If some point $Y$ of $S$ were not covered by this circle, then the distance $XY$ would exceed $1/\sqrt{3}$ and we would have the contradiction that $X$ would not lie in the circle of the collection which is drawn about the point $Y$.

Helly's theorem is the key to many surprising results. Two of its noteworthy corollaries are given in the exercises.

Now let us turn to our second application, in which we shall establish a property that I consider to be one of the most amazing things in all of geometry!!

The setting is a finite set of $n$ points in the plane ($n$ arbitrary). A straight line drawn across the set splits it into two parts; and, in counting up the number of points in each part, let the points on the line itself be counted for each side. Then we have the following incredible result.

*For each set S of n points in the plane, there exists a point O in the plane (not necessarily in S, itself) having the property that every straight line through O has at least one-third of the points on each side of it.*

A finite set of points is necessarily bounded, and for any bounded figure $S$ (perhaps in many pieces), it is similarly true that there exists a point $I$ such that every straight line through $I$ has on each side at least one-third of the perimeter of $S$, and also a point $J$ with the same

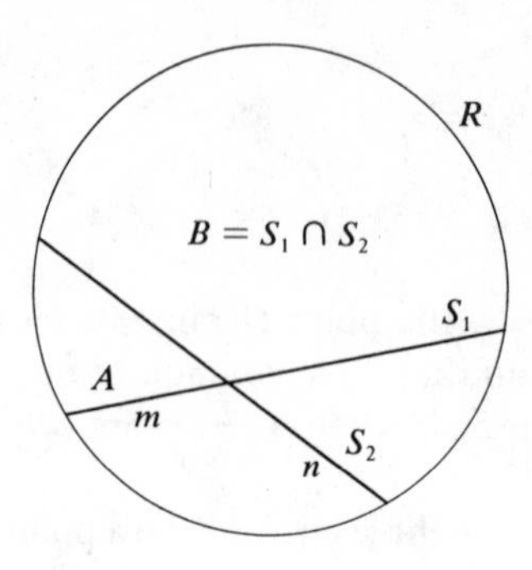


Figure 4

property concerning the area of $S$. Since these three theorems have almost identical proofs, we shall consider in detail only the first one.

As just noted, because $S$ is finite, it must be bounded. Accordingly, let $R$ be a circle which contains $S$ in its interior. A straight line $m$ that cuts across $S$ will divide $R$ into two circular segments (Figure 4). Let us take these segments to be closed sets; in particular, then, a segment is considered to include its bounding chord.

Now let us consider the infinite collection $C$ of all the segments of $R$ which contain more than two-thirds of the points of $S$ (i.e., more than $2n/3$ of the points). (There is an infinity of such segments, all with parallel edges, for each direction in the plane. $C$ contains all the segments in all directions.)

With the infinite case of Helly's theorem we shall show that there exists a point $O$ which lies in all the members of $C$ (which are obviously convex and bounded).

To this end, consider any three segments $S_1$, $S_2$, $S_3$ of $C$. Suppose that the bounding edges of $S_1$ and $S_2$ are $m$ and $n$, respectively. In general, $n$ cuts across $S_1$, separating a part $A$ and the intersection $B = S_1 \cap S_2$ (Figure 4). Since $S_2$ contains more than $2n/3$ of the points of $S$, the part $A$, lying outside $S_2$, cannot contain as many as $n/3$ of the points. But $S_1$ contains more than $2n/3$ of the points. Thus there must be more than $n/3$ of them in $B$. Consequently, turning to $S_3$, even if $S_3$ were to claim all the points of $S$ that are not in $B$, there would still not be enough to make up its quota of more than two-thirds of the points. Thus $S_3$ must pick up at least one point of $S$ from

$B$, implying that $S_1$, $S_2$, $S_3$ have at least one point in common. By Helly's theorem, then, there is a point $O$ that is common to all the segments in $C$.

The proof is easily completed by showing that every straight line $q$ through $O$ has at least $n/3$ of the points of $S$ on each side (counting the ones on $q$, itself). Suppose to the contrary that the number of points of $S$ which occur on $q$ and to one side, say the left, is $< n/3$. The rest of $S$, a subset $T$ which consists of more than $2n/3$ of the points, must occur strictly in the interior of the other side of $q$ (Figure 5). Since $T$ is finite, it must have some point $Y$ which is closest to $q$ (perhaps more than one). Then a straight line $r$ parallel to $q$, which lies between $Y$ and $q$, has on its right side the entire subset $T$ (with its more than $2n/3$ of the points of $s$), and therefore the right side of $r$ determines a segment that belongs to the collection $C$. However, this segment clearly fails to contain the special point $O$, in contradiction to the above result of Helly's theorem.

It is easy to see that no number greater than $n/3$ can be used in this theorem. If the points of $S$ are divided equitably among small regions near the vertices of a triangle $ABC$, it is clear that no point $O$ exists for which *every* line through $O$ has *more* than $n/3$ of the points on each side (Figure 6).

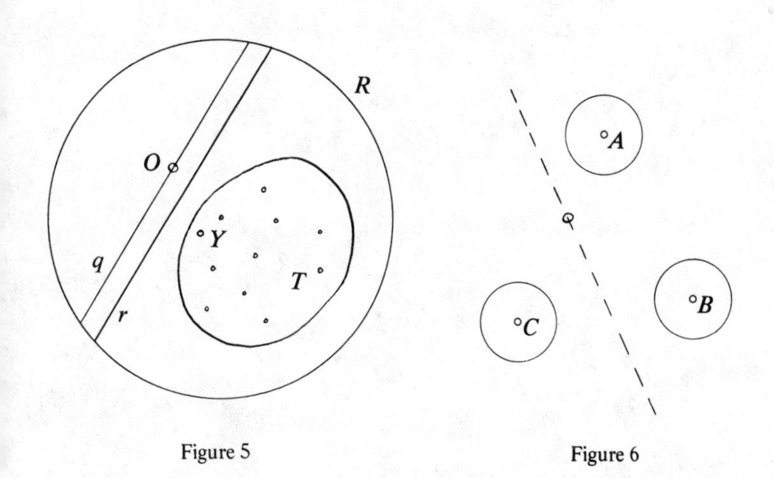


Figure 5

Figure 6

**Exercises**

1. If for each three segments of a given set of $n$ parallel (straight) segments, there exists a straight line which crosses the three segments, prove that there exists a straight line which crosses all $n$ of the segments.

2. Prove that inside every bounded convex figure there exists a point $O$ which lies in the middle third of every chord through it.

CHAPTER **14**

# AN INTRODUCTION TO RAMANUJAN'S HIGHLY COMPOSITE NUMBERS

Much has been written about the great self-taught Indian mathematician Srinavasa Ramanujan (1887–1920). His story is told briefly in J. R. Newman's monumental *The World of Mathematics* (Vol. 1, pages 366–376); [**1**] is a full biography, and [**2**] and [**3**] contain many of his contributions to mathematics. In [**2**] there is a long paper on what Ramanujan calls "highly composite numbers." In this introduction, we shall establish only one simple theorem concerning the form such numbers must take.

As one might expect from the name, these numbers have particularly many divisors. While lots of integers have many divisors, a highly composite number must have more divisors than any smaller positive integer has. Employing $d(N)$ to denote the number of (positive) divisors of the positive integer $N$, then $N$ is highly composite if and only if

$$d(N) > d(N') \quad \text{for all } N' < N.$$

These numbers may be contrasted with the prime numbers, which have few divisors. From Table 1 we see that the first six highly composite numbers are 2, 4, 6, 12, 24, and 36. There are 25 highly composite numbers between 2 and 50,000, and in his paper, Ramanujan gives a list of the more than 100 highly composite numbers up to the number 6746328388800. Here are a few samples and their prime decompositions:

$$5040 = 2^4 \cdot 3^2 \cdot 5 \cdot 7$$

$$332640 = 2^5 \cdot 3^3 \cdot 5 \cdot 7 \cdot 11$$

$$1081080 = 2^3 \cdot 3^3 \cdot 5 \cdot 7 \cdot 11 \cdot 13$$

$$43243200 = 2^6 \cdot 3^3 \cdot 5^2 \cdot 7 \cdot 11 \cdot 13$$

$$27935107200 = 2^7 \cdot 3^3 \cdot 5^2 \cdot 7 \cdot 11 \cdot 13 \cdot 17 \cdot 19$$

$$2248776129600 = 2^6 \cdot 3^3 \cdot 5^2 \cdot 7^2 \cdot 11 \cdot 13 \cdot 17 \cdot 19 \cdot 23$$

His list represents a staggering amount of calculation, for he knew of no formula for generating them. This brings to mind a penetrating remark made by Harold Edwards in the preface of his book *Fermat's Last Theorem* (Springer-Verlag Graduate Texts Series, number 50):

> ... as even a superficial glance at history shows, Kummer and the other great innovators in number theory did vast amounts of computation and gained much of their insight in this way. I deplore the fact that contemporary mathematical education tends to give students the idea that computation is demeaning drudgery to be avoided at all costs.

For the sake of completeness, it should be noted that Ramanujan evidently missed one highly composite number in his list, for 29331862500 does not appear.

From a study of Ramanujan's list and their prime decompositions, it is a short step to conjecture the following three engaging characteristics of highly composite numbers, whose proof will constitute the remainder of our discussion:

THEOREM. *If $N = 2^{a_2}3^{a_3}\cdots p^{a_p}$ is the prime decomposition of a highly composite number N, then*

(i) *the primes 2, 3, ..., p form an unbroken string of consecutive primes as far as they go,*
(ii) *the exponents are nonincreasing: $a_2 \geq a_3 \geq \cdots \geq a_p$,*
(iii) *the final exponent $a_p$ is always 1, except for the two cases $N = 4 = 2^2$ and $N = 36 = 2^2 \cdot 3^2$, when it is 2.*

| $N$ | 1 | 2 | 3 | 4 | 5 | 6 | 7 | 8 | 9 | 10 | 11 | 12 | 13 | 14 | 15 | 16 | 17 | 18 | 19 |
|---|---|---|---|---|---|---|---|---|---|---|---|---|---|---|---|---|---|---|---|
| $d(N)$ | 1 | 2 | 2 | 3 | 2 | 4 | 2 | 4 | 3 | 4 | 2 | 6 | 2 | 4 | 4 | 5 | 2 | 6 | 2 |

| $N$ | 20 | 21 | 22 | 23 | 24 | 25 | 26 | 27 | 28 | 29 | 30 | 31 | 32 | 33 | 34 | 35 | 36 | 37 | 38 ··· |
|---|---|---|---|---|---|---|---|---|---|---|---|---|---|---|---|---|---|---|---|
| $d(N)$ | 6 | 4 | 4 | 2 | 8 | 3 | 4 | 4 | 6 | 2 | 8 | 2 | 6 | 4 | 4 | 4 | 9 | 2 | 4 ··· |

Table 1

Parts (i) and (ii) are particularly easy to prove from a simple formula for $d(N)$, which we shall begin by deriving. If $N = p_1^{a_1}p_2^{a_2}\cdots p_n^{a_n}$ is the prime decomposition of a positive integer $N$, then every divisor $m$ of $N$ must have a like prime decomposition

$$m = p_1^{b_1}p_2^{b_2}\cdots p_n^{b_n},$$

where each $b_i$ is restricted to the range $(0, 1, 2, \ldots, a_i)$. Thus, in constructing a divisor of $N$, there are $a_1 + 1$ choices for the value $b_1$, $a_2 + 1$ choices for $b_2$, and so on, implying that the *number* of divisors is

$$d(N) = (a_1 + 1)(a_2 + 1)\cdots(a_n + 1).$$

Observe that $d(N)$ depends only on the collection of exponents, not on the primes themselves, and that the value of $d(N)$ is not changed by permuting the exponents over the same or a different set of prime bases.

(i) Suppose that some prime number $P$ is missing from the string of primes $2, 3, \ldots, p$ in the decomposition of the highly composite number $N$:

$$N = 2^{a_2}3^{a_3}\cdots(\text{missing a factor of } m^r)\cdots p^{a_p}.$$

Dropping the final factor $p^{a_p}$ and including the smaller $m^{a_p}$ instead, we obtain a smaller number

$$N' = 2^{a_2}3^{a_3}\cdots(m^{a_p})\cdots$$

having the same exponents and therefore the same value of $d$:

$$N' < N \text{ and } d(N') = d(N).$$

This contradicts the highly composite character of $N$, and part (i) follows.

(ii) Part (ii) is quickly established in a similar way. Suppose, contrary to desire, that some exponent $m + n$ exceeds an earlier exponent $m$:

$$N = 2^{a_2}3^{a_3} \cdots p_1^m \cdots p_2^{m+n} \cdots p^{a_p}.$$

Interchanging the exponents of $p_1$ and $p_2$, then, we obtain the contradiction of the smaller number

$$N' = 2^{a_2}3^{a_3} \cdots p_1^{m+n} \cdots p_2^{m} \cdots p^{a_p}$$

having $d(N') = d(N)$.

(iii) Before we are finished with part (iii), we shall need the following results from number theory. Suppose $p_1, p_2, p_3, p_4, p_5$ are consecutive in the sequence of prime numbers. Then

(a) $p_1^2 > p_2$ for all choices of $p_1$,

(b) $p_1^3 > p_4$ for all $p_1$,

(c) $p_2p_3 > p_1p_4$ for $5 \leq p_3 \leq 19$,

(d) $p_1p_2p_3 > p_4p_5$ for $p_3 \geq 11$.

There is no difficulty in verifying part (c) by direct observation of the few cases it describes, and the other parts follow easily from the very useful theorem known as Bertrand's Postulate. This declares the existence of a prime number between $n$ and $2n$ for $n$ a positive integer $> 1$. A complete account of Paul Erdös' marvellous elementary proof of this difficult theorem is given in one of my all-time favorite books—Waclaw Sierpiński's *Theory of Numbers* [4] (see the chapter on prime numbers). Erdös did this when he was 17 or 18 (1931); an auspicious beginning to a truly amazing creative career which is still flourishing today.

By Bertrand's Postulate we have $p_{n+1} < 2p_n$, and repeated application gives $p_{n+1} < 2p_n < 4p_{n-1} < 8p_{n-2} < \cdots$. Any of (a), (b), (d) can be verified directly for the small primes and Bertrand's Postulate takes care of all cases beyond a very small minimum:

(a) $p_2 < 2p_1 < p_1^2$ for $p_1 \geq 3$;

(b) $p_4 < 2p_3 < 4p_2 < 8p_1 < p_1^3$ for $p_1^2 > 8$, i.e., $p_1 \geq 3$;

(d) $p_4 < 2p_3$, and $p_5 < 2p_4 < 4p_3 < 8p_2$;

thus

$$p_4 p_5 < 16 p_2 p_3 < p_1 p_2 p_3 \quad \text{for } p_1 \geq 17.$$

The arguments that follow bear a considerable resemblance to one another, but they are so simple and straightforward that I hope you will breeze through them without tedium.

The first use we make of these results is to show that the final exponent cannot exceed 2. Suppose, to the contrary, that the final prime $p_n$ has exponent $m + 2$ $(m \geq 1)$:

$$N = 2^{a_2} 3^{a_3} \cdots p_n^{m+2}.$$

By (a), we have $p_n^2 > p_{n+1}$, and so the number

$$N' = \left(\frac{p_{n+1}}{p_n^2}\right) \cdot N = 2^{a_2} 3^{a_3} \cdots p_n^m p_{n+1} < N.$$

Since $N$ is highly composite, we have $d(N') < d(N)$, that is,

$$(a_2 + 1)(a_3 + 1) \cdots (m + 1)2 < (a_2 + 1)(a_3 + 1) \cdots (m + 3)$$

$$(m + 1)2 < m + 3$$

$$m < 1,$$

a contradiction.

Next we similarly prove that the "third last" exponent cannot exceed 4. Suppose that the last three primes are $p_1$, $p_2$, $p_3$ (they must be consecutive primes by (i)), and that the exponent of $p_1$ is $4 + m$ $(m \geq 1)$:

$$N = 2^{a_2} 3^{a_3} \cdots p_1^{4+m} p_2^s p_3^t.$$

By (b), we have $p_1^3 > p_4$, implying that

$$N' = 2^{a_2} 3^{a_3} \cdots p_1^{1+m} p_2^s p_3^t p_4 < N,$$

which leads to $d(N') < d(N)$,

$$(a_2 + 1) \cdots (2 + m)(s + 1)(t + 1)2$$
$$< (a_2 + 1) \cdots (5 + m)(s + 1)(t + 1)$$
$$(2 + m)2 < 5 + m$$
$$m < 1,$$

a contradiction.

Now let us consider highly composite numbers $N$ which have at least three prime divisors, and suppose that the last three primes in the prime decomposition of such an $N$ are $p_1$, $p_2$, and $p_3$. We show that the final exponent is 1 first in the case $p_3 \leq 19$, and then for the somewhat overlapping case of all $p_3 \geq 11$.

**$\mathbf{p_3 \leq 19}$.** We have seen that the final exponent cannot exceed 2. Suppose, then, that it is 2:

$$N = 2^{a_2} 3^{a_3} \cdots p_1^r p_2^s p_3^2.$$

With two primes smaller than $p_3$, $p_3$ must be at least 5, and for $p_3 \leq 19$, by (c) we have that $p_2 p_3 > p_1 p_4$. Consequently

$$N' = \left(\frac{p_1 p_4}{p_2 p_3}\right) \cdot N = 2^{a_2} 3^{a_3} \cdots p_1^{r+1} p_2^{s-1} p_3 p_4 < N,$$

and we have $d(N') < d(N)$, yielding

$$(r + 2) \cdot s \cdot 2 \cdot 2 < (r + 1)(s + 1) \cdot 3,$$

which reduces directly to

$$r(s - 3) + 5s < 3. \tag{1}$$

Now, since the exponents are nonincreasing, we have $r \geq s \geq 2$. Clearly, for $s \geq 3$, the value of $r(s - 3) + 5s$ is at least 15; and for $s = 2$, relation (1) becomes $7 < r$. However, we have seen that the third last exponent cannot exceed 4, and we have a contradiction in all cases.

**$\mathbf{p_3 \geq 11}$.** Again suppose that the final exponent is 2:

$$N = 2^{a_2} 3^{a_3} \cdots p_1^r p_2^s p_3^2.$$

By (d), we have

$$N' = \left(\frac{p_4p_5}{p_1p_2p_3}\right)\cdot N = 2^{a_2}3^{a_3}\cdots p_1^{r-1}p_2^{s-1}p_3p_4p_5 < N,$$

and $d(N') < d(N)$ yields

$$rs\cdot 2\cdot 2\cdot 2 < (r+1)(s+1)\cdot 3$$

$$8 < \left(1+\frac{1}{r}\right)\left(1+\frac{1}{s}\right)\cdot 3.$$

Since $r \geq s \geq 2$, we have each of $1 + (1/r)$ and $1 + (1/s)$ is at most $1 + (1/2)$. Thus

$$\left(1+\frac{1}{r}\right)\left(1+\frac{1}{s}\right)\cdot 3 \leq \frac{3}{2}\cdot\frac{3}{2}\cdot 3 = \frac{27}{4} < 8,$$

a contradiction. Therefore, if $N$ has three or more prime divisors, property (iii) is valid, that is, the final exponent is 1.

It remains to consider $N$ which have only one or two prime divisors. Since the prime divisors form an unbroken string in the sequence of primes, we have only numbers of the form

$$N = 2^a3^b.$$

From our earlier results we have $a \geq b$ and $b \leq 2$, and we are interested only in the exceptional cases when $b$ is not 1.

For $b = 0$, we have $N = 2^a$, where $a \leq 2$. Thus $a = 2$ gives the only exceptional case, and we get $N = 2^2 = 4$.

For $b = 2$, we have $N = 2^a3^2$. Since $a \geq 2$, suppose $a = 2 + n$ $(n \geq 0)$. Since $2\cdot 3 > 5$, we have

$$N = 2^{2+n}3^2 > 2^{1+n}\cdot 3\cdot 5 = N',$$

and then

$$d(N') < d(N),$$

giving

$$(2+n)\cdot 2\cdot 2 < (3+n)\cdot 3,$$

and

$$n < 1.$$

This implies that $n = 0$ and $N = 2^2 3^2 = 36$.

**References**

1. S. R. Ranganathan, Ramanujan: The Man and the Mathematician, Asia Publishing House, London, 1967.
2. G. H. Hardy, Aiyar, and Wilson, Collected Papers of Ramanujan, Chelsea Publishing Co., New York, 1927.
3. G. H. Hardy, Ramanujan: 12 Lectures on his Life and Work, Chelsea Publishing Co., New York, 1959.
4. W. Sierpiński, Theory of Numbers, Monographs in Mathematics, Volume 42, Panstwowe Wydawnictwo Naukowe, Warsaw, 1964.

CHAPTER **15**

# ON SETS OF POINTS IN THE PLANE

## 1. Sets which have a Center

(a) Since the areas of two semicircles of a disk are the same, one can hardly object to the statement that a diameter divides a disk "in half." However, when it comes to an equitable division of the *points* of a disk (considered to consist of the interior and the bounding circumference), one runs into trouble trying to make any two "halves" which are congruent. Surprisingly, it is simply impossible to color each point of a disk either red or blue in such a way that the resulting sets $R$ and $B$ (for red and blue) are congruent. This is quite easy to see indirectly.

Suppose there were to be a way to color the disk $D$ so that $R \equiv B$. The center $O$ would have to belong to one of these sets, say $R$. In a copy $D'$ of $D$, let the colored sets be denoted by $R'$ and $B'$. Because $R \equiv B$, $D'$ can be placed on top of $D$ so that the blue set $B'$ occurs precisely on top of the red set $R$ (Figure 1). In particular, the red center $O$ in $R$ must be covered by some blue point $O'$. Since $O'$ is blue, it cannot be the center of $D'$ (which is a copy of the red center $O$). Therefore $D'$ overhangs $D$, producing two lunes $X$ and $Y$, as shown in Figure 1.

Since all the blue points of $D'$ are busy covering red points, the entire unengaged lune $X$ must consist of red points. Similarly, since all the red points of $D$ are underneath $D'$, the uncovered lune $Y$ must be entirely blue. Now, no matter how little $D'$ and $D$ might be off-center, the length of the arc around each of the lunes $X$ and $Y$ must exceed a semicircle. Therefore, each of the colors contains a continuous arc exceeding a semicircle, implying the contradiction that some point on the circumference belongs to both sets.

It seems pretty clear that the *center* is the fly in the ointment, "un-

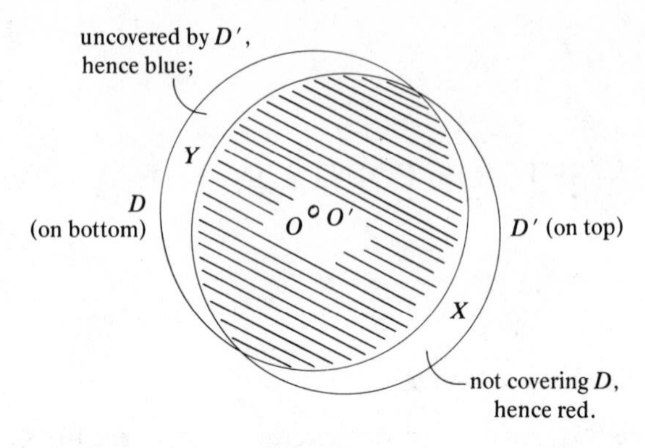


Figure 1

balancing" the set into which it is placed when the points are divided up. Of course, if the given set of points $S$ extends across the plane infinitely far, the desired partitioning into congruent subsets might be possible even though $S$ contains a center. For example, the integer points on the $x$-axis have center the origin (in fact, this set is symmetric about each of its points), and yet two congruent subsets result from coloring every second point red and the rest blue, as a unit translation shows. However, it must be different for *bounded* sets. So long as a bounded set of points has a center that belongs to the set, surely the sets $R$ and $B$ must fail to be congruent. It is not amazing that this is simply not true!! In part (b) we shall construct and "halve" a bounded set $S$ that contains its center.

(b) Our discussion takes place in the complex plane, where the point $(x, y)$ is represented by the number $x + yi$ ($i = \sqrt{-1}$) and is alternatively thought of as the vector from the origin to the point $(x, y)$.

First we construct a preliminary set of points. Let the infinite set of points given by

$$z_n = e^{in}, \qquad n = 0, 1, 2, \ldots,$$

be marked around the unit circle, center the origin. Geometrically these are obtained by dividing the circumference into arcs of length equal to the radius, beginning at (1,0), and proceeding counterclockwise (Figure 2).

The sets $R$ and $B$ of $S$ are derived from these points as follows. To get $B$, drop the point (1,0) from the set and translate all the other $z_n$ one unit in the direction of the negative $x$-axis (Figure 3):

$$B \equiv z_n - 1, \qquad n = 1, 2, 3, \ldots.$$

To get $R$, all the $z_n$ are first reflected in the origin (to yield $-z_n$) and then translated one unit in the direction of the positive $x$-axis (Figure 3):

$$R \equiv 1 - z_n, \qquad n = 0, 1, 2, 3, \ldots.$$

Observe that, because (1,0) is dropped in making $B$, the range of $n$, which begins at 0 for $R$, begins only at 1 for $B$. Accordingly, $B$ does *not* contain the origin, while $R$ does (as the final image of (1,0)). Thus the sets $R$ and $B$ are disjoint, occurring as they do on circles whose only common point does not belong to $B$. However, the set $S$,

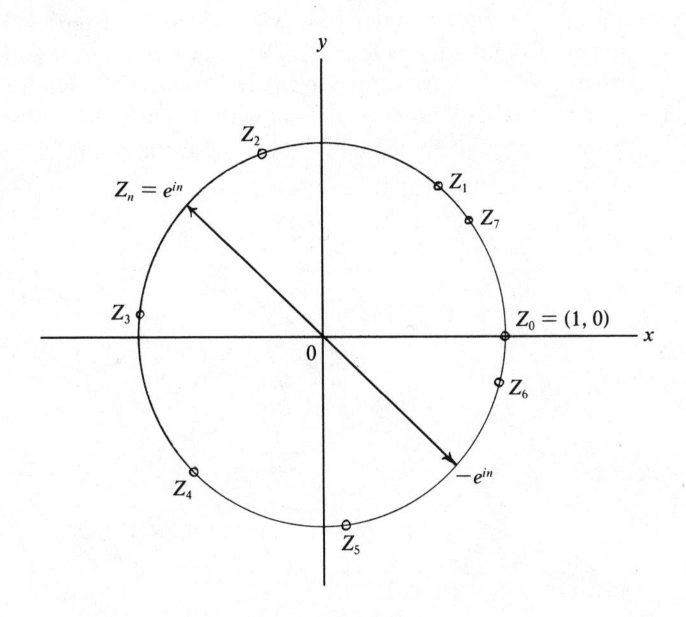


Figure 2

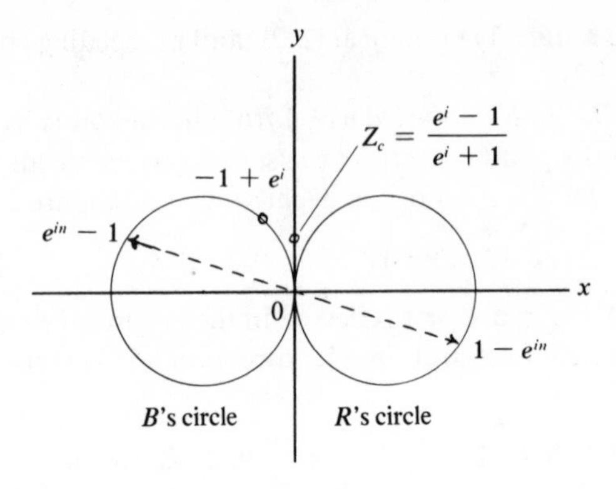


Figure 3

consisting of the union of $R$ and $B$, is clearly symmetric about the origin: opposite each point $z_n - 1$ of $B$ ($n = 1, 2, 3, \ldots$) is the corresponding point $1 - z_n$ of $R$. In summary, then, we have a bounded set $S$, having its center (the origin) among its points, divided into two disjoint subsets $R$ and $B$. We now show that $R$ and $B$ are congruent. We shall demonstrate the congruence by defining a rigid motion that takes $R$ into coincidence with $B$. The motion is simply the rotation through an angle of $\pi + 1$ radians about the center

$$z_c = \frac{e^i - 1}{e^i + 1}.$$

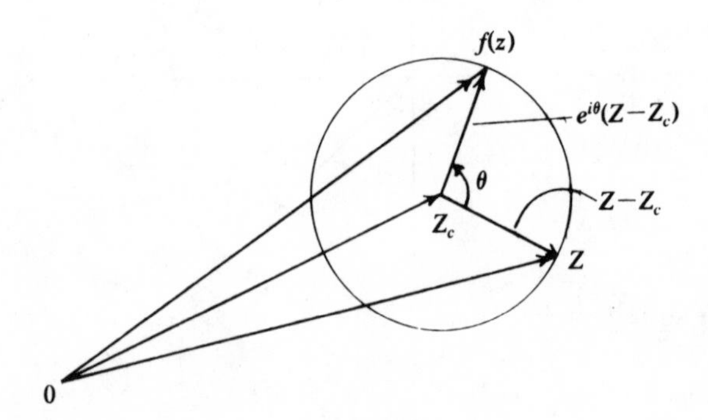


Figure 4

We recall that the rotation about center $z_c$ through an angle $\Theta$ carries the point $z$ into the point $f(z)$ given by $f(z) = z_c + e^{i\Theta}(z - z_c)$ (Figure 4). Accordingly, the rotation in question is given by

$$f(z) = \frac{e^i - 1}{e^i + 1} + e^{i(\pi+1)}\left(z - \frac{e^i - 1}{e^i + 1}\right).$$

Since $e^{\pi i} = -1$, we have $e^{i(\pi+1)} = -e^i$. Therefore

$$f(z) = \left(\frac{e^i - 1}{e^i + 1}\right) - e^i z + e^i\left(\frac{e^i - 1}{e^i + 1}\right)$$

$$= e^i - 1 - e^i z,$$

or

$$f(z) = -1 + e^i(1 - z).$$

The effect of this rotation on the points $z = 1 - z_n$ of $R$ is to carry them to the images

$$f(z) = f(1 - z_n) = -1 + e^i(z_n) \qquad (\text{recall } z_n = e^{in})$$

$$= -1 + e^{i(n+1)}.$$

For $n = 0, 1, 2, \ldots$, this is precisely the set of points $B = -1 + e^{in}$, where $n$ takes the range $n = 1, 2, 3, \ldots$.

Intuitively we feel that the center $z_c$ of this rotation should lie on the $y$-axis. This is correct, and its easy proof is left to the reader.

In constructing the set $R$, the reflection in the origin has the effect of starting the counterclockwise parade of points $z_n$ around the unit circle from the point $(-1, 0)$ instead of $(1, 0)$ (the direction is still counterclockwise). The rotation just described carries the base-circle of $R$ onto that of $B$ and the center $(0, 0)$ of $S$ onto the point $-1 + e^i$ on $B$'s circle; from there the points of $R$ and $B$ march around the circle together (Figure 3).

## 2. A Pathological Set

Although it is impossible to divide the points of a disk into two congruent subsets, there exist any number of sets for which this can be done. The union of any two disjoint congruent sets $R$ and $B$ trivially suffices. Each of $R$ and $B$ constitute what might legitimately be called "half" of the set, and everybody knows that half of something is not the same as all of it. As Euclid said, "The whole is greater than the part." What an incredible surprise, then, it is to learn that

> there exists in the plane a set of points $S$ which is the union of two disjoint congruent *proper* subsets $R$ and $B$, where each of $R$ and $B$ is also congruent to the entire set $S$, itself.

In order to describe the set we again turn to the complex plane. The complex number

$$z = a_n e^{in} + a_{n-1} e^{i(n-1)} + \cdots + a_1 e^i + a_0,$$

where $n$ and the coefficients $a_i$ are nonnegative integers, describes the position of some point in the plane. As $n$ and the $a_i$ range over all possible combinations of nonnegative integers, a tremendous infinite set of points $S$ is produced. We can divide the points $z$ of $S$ into two subsets $R$ and $B$ according to the value of the absolute term $a_0$:

$$\begin{aligned} &\text{if} \quad a_0 = 0, \quad \text{put } z \text{ in } R; \\ &\text{if} \quad a_0 > 0, \quad \text{put } z \text{ in } B. \end{aligned}$$

Assuming for the moment that each point $z$ has only one representation in the prescribed form, no point is put into both $R$ and $B$, implying that $R$ and $B$ are disjoint. (We shall return to the uniqueness of this expression at the end of the discussion.) We will establish our claim by showing the amazing fact that each of $R$ and $B$ is congruent to $S$ itself (from which it follows that they are also congruent to each other).

In the case of $S$ and $B$, let $S$ be translated one unit in the direction of the positive $x$-axis. This takes the point $z$ into $z + 1$, and since $a_0 \geq 0$ for every point $z$ in $S$, the absolute term in $z + 1$ is strictly positive. Consequently, the image point $z + 1$ is always a point of $B$. Conversely, each point $z$ of $B$ has $a_0 > 0$, making $a_0 - 1$ nonnegative, implying that $z - 1$ belongs to the set $S$. Under the translation, this point $z - 1$ slides over to cover the point $z$ of $B$. Thus the translation leaves no point of $B$ uncovered, and carries every point of $S$ to an image in $B$. Therefore $S \equiv B$.

To show that $S$ and $R$ are congruent, let $S$ be rotated about the origin through an angle of one radian. This transformation carries the point $z$ of $S$ into the point $ze^i$:

$$ze^i = a_n e^{i(n+1)} + a_{n-1}e^{in} + \cdots + a_1 e^{2i} + a_0 e^i.$$

Having absolute term equal to 0, this image point belongs to $R$. Conversely, each point of $R$, having $a_0 = 0$, is of the form

$$z = a_n e^{in} + a_{n-1}e^{i(n-1)} + \cdots + a_1 e^i,$$

implying that the point

$$\frac{z}{e^i} = a_n e^{i(n-1)} + a_{n-1}e^{i(n-2)} + \cdots + a_1$$

is a point of $S$. Clearly the rotation moves this point of $S$ to cover the point $z$ of $R$, and we conclude that $S \equiv R$.

It remains only to settle the question of the uniqueness of the expression for $z$. This is a difficult area, whose proofs are far beyond the scope of the present discussion. The desired result follows from the transcendental character of the number $e^i$ (that is to say, $e^i$ is *not* a root of any polynomial equation which has integral coefficients). It was only in 1873 that the great French mathematician Charles Hermite proved that $e$ is transcendental. And it does not follow automatically from the transcendence of $e$ that $e^i$ is also transcendental, for it is well known that the much more likely $e^{\pi i}$ is just the integer $-1$. For more information on this matter the reader is referred to Theorem 1.4 on page 6 of Alan Baker's authoritative *Transcendental Number Theory* (Cambridge University Press, 1979). On the assumption of the transcendence of $e^i$, we can reach our goal indirectly in one step.

Suppose some point $z$ were to be specified by two different expressions

$$z = a_n e^{in} + \cdots + a_0 = b_m e^{im} + \cdots + b_0.$$

By simply transposing, we obtain a relation that displays $e^i$ as a root of a polynomial equation which has integral coefficients. With this contradiction, the argument is complete.

**References**

1. Hadwiger, Debrunner, and Klee, Combinatorial Geometry in the Plane, Holt, Rinehart, and Winston, New York, 1964.
2. H. Meschkowski, Unsolved and Unsolvable Problems in Geometry, Oliver and Boyd, Edinburgh and London, 1966.

CHAPTER **16**

# TWO SURPRISES FROM ALGEBRA

## 1. A Gem from the East

The following remarkable theorem made its way to Waterloo among the souvenirs of a recent trip to India by my colleague Bruce Richmond.

THEOREM. *Let $f(x)$ be a monic polynomial having integral coefficients. Then, on the interval $-2 \le x \le 2$, the function must attain a magnitude of at least 2: i.e.,*

$$\max_{-2\le x\le 2} |f(x)| \ge 2.$$

It was presented to Bruce by the mathematicians Balasubramanian and Ramachandra of the Tata Institute in Bombay, but I am told that the result is due to the great Russian mathematician Chebyshev. It should come as no surprise, then, that the key role in our proof is played by the Chebyshev polynomials. Let us begin with a brief introduction to these functions.

(i) The Chebyshev polynomials constitute a sequence of monic polynomials having integral coefficients, and they are defined recursively by

$$f_0(x) = 2, \qquad f_1(x) = x,$$

and, for $n \ge 2$,

$$f_n(x) = xf_{n-1}(x) - f_{n-2}(x).$$

This yields the sequence $2, x, x^2 - 2, x^3 - 3x, x^4 - 4x^2 + 2, \ldots,$ where it is clear that $f_n(x)$ has degree $n$. For our purposes, the important property of these polynomials is that, for all $n$,

$$f_n\left(y + \frac{1}{y}\right) = y^n + \frac{1}{y^n}.$$

This follows immediately by induction and the easy proof is left to the reader.

(ii) For $x$ in the range $-2 \leq x \leq 2$, we have $-1 \leq x/2 \leq 1$, and $x/2$ is the cosine of some angle $\Theta$:

$$\frac{x}{2} = \cos\Theta, \qquad \Theta = \cos^{-1}\frac{x}{2}, \qquad x = 2\cos\Theta = e^{i\Theta} + e^{-i\Theta}.$$

Then, for $x$ in the range $[-2, 2]$, we have, for all $n$, that the Chebyshev polynomials are given by

$$f_n(x) = f_n(e^{i\Theta} + e^{-i\Theta}) = e^{in\Theta} + e^{-in\Theta} = 2\cos n\Theta,$$

that is,

$$f_n(x) = 2\cos\left[n\cos^{-1}\frac{x}{2}\right].$$

Turning to the proof, then, let $n$ be the degree of the given polynomial $f(x)$. Since both $f(x)$ and $f_n(x)$ are monic polynomials of degree $n$, their difference

$$F(x) = f(x) - f_n(x)$$

is a polynomial of degree not exceeding $n - 1$. Accordingly, $F(x)$ cannot have more than $n - 1$ zeros without being identically zero. That is to say, if the graphs of $f(x)$ and $f_n(x)$ intersect more than $n - 1$ times, we have

$$f(x) \equiv f_n(x).$$

Now we proceed indirectly: suppose that $|f(x)| < 2$ for every $x$ in the range $[-2, 2]$, that is, the graph of $f(x)$ lies strictly between the lines $y = 2$ and $y = -2$ for $x$ in this range.

Now,

| as | $x$ | goes from | $-2$ to $2$, |
|---|---|---|---|
| | $\frac{x}{2}$ | goes from | $-1$ to $1$, |

$$\cos^{-1}\frac{x}{2} \quad \text{goes from} \quad \pi \text{ to } 0,$$

$$n\cos^{-1}\frac{x}{2} \quad \text{goes from} \quad n\pi \text{ to } 0,$$

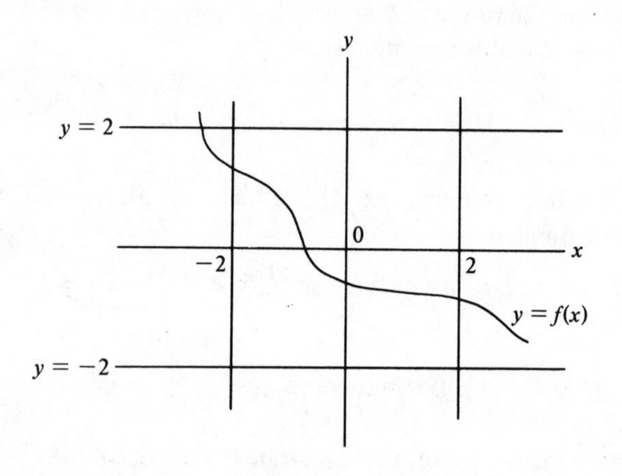


and the graph of $f_n(x) = 2\cos[n\cos^{-1}(x/2)]$ runs across the strip between the lines $y = 2$ and $y = -2$ $n$ times, intersecting the (continuous) graph of $f(x)$ at least once each time, for a total of at least $n$ intersections altogether. Thus we must have $f(x) \equiv f_n(x)$. But, in this case, we have

$$f(2) = f_n(2) = f_n\left(1 + \frac{1}{1}\right) = 1^n + \frac{1}{1^n} = 2,$$

contradicting $|f(2)| < 2$, and the theorem follows.

## 2. Series Multisection

Finding the sum of a series has been a major occupation among mathematicians since time immemorial. One may well get the feeling of having ventured into virgin territory, however, when attempting to find the likes of the sum of every seventh term, beginning at the fifth one. Our second surprise in this chapter is a remarkable formula, which can be established with nothing more than freshman mathe-

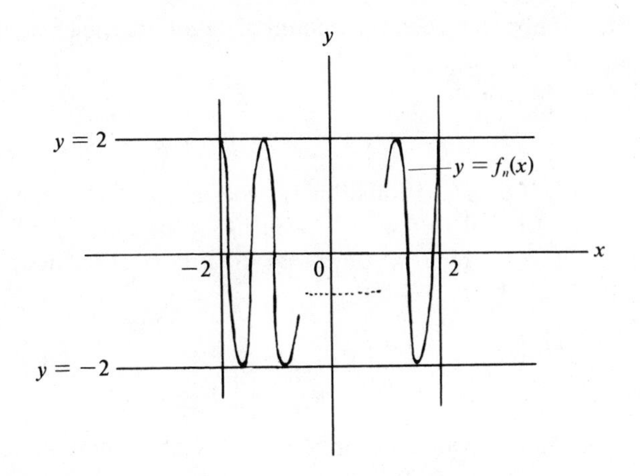


matics, that gives the sum of every $n$th term of a series, beginning with the $j$th one, $j \leq n$, which we denote by $S(n, j)$:

if

$$f(x) = f_0 + f_1 x + f_2 x^2 + \cdots + f_n x^n + \cdots$$

is a finite or infinite series, then

$$S(n, j) = f_j x^j + f_{j+n} x^{j+n} + f_{j+2n} x^{j+2n} + \cdots$$

is given by the formula

$$S(n, j) = \frac{1}{n} \sum_{t=0}^{n-1} w^{-jt} f(w^t x), \tag{1}$$

where $w = e^{2\pi i/n}$, an $n$th root of unity.

The right-hand side of (1) is an unsimplified power series in $x$ (each of its $n$ summands contains a power series $f(w^t x)$), and the proof of the equation consists simply in showing that, for every $k$, there are the same number of $x^k$ on each side; in particular, we need to show that, as $k$ runs through 0, 1, 2, ...,

*whenever $k \equiv j \pmod{n}$, the coefficient $c_k$ of $x^k$ is $f_k$, and is zero otherwise.*

Although this involves a certain amount of detail, it is just a matter of checking calculations.

In the power series

$$f(w^t x) = f_0 + f_1 w^t x + \cdots + f_k (w^t x)^k + \cdots,$$

the coefficient of $x^k$ is $f_k w^{tk}$, making the coefficient of $x^k$ in the quantity $w^{-jt} f(w^t x)$ equal to $f_k w^{t(k-j)}$. Adding up over $t = 0, 1, \ldots, n-1$, and dividing by $n$, we have that the final coefficient of $x^k$ on the right side of (1) is

$$c_k = \frac{1}{n} \sum_{t=0}^{n-1} f_k w^{t(k-j)}. \tag{2}$$

Now, $j$ is a given constant. Therefore, specifying $k$ determines the number $r = k - j$, and (2) yields

$$c_k = \frac{f_k}{n} \sum_{t=0}^{n-1} w^{rt} = \frac{f_k}{n} (1 + w^r + w^{2r} + \cdots + w^{(n-1)r}). \tag{3}$$

Whenever $k \equiv j \pmod{n}$, we have $r = k - j = mn$, a multiple of $n$. In this case (observing that $w^n = 1$),

$$w^r = w^{mn} = (w^n)^m = 1^m = 1,$$

and (3) gives

$$c_k = \frac{f_k}{n} (1 + 1 + 1 + \cdots + 1) = \frac{f_k}{n} \cdot n = f_k,$$

as desired.

Now, for all values of $r$, we have

$$(w^r)^n - 1 = (w^n)^r - 1 = 1^r - 1 = 0,$$

which factors to give

$$(w^r - 1)(1 + w^r + w^{2r} + \cdots + w^{(n-1)r}) = 0. \tag{4}$$

Finally, then, consider the times when $k \not\equiv j \pmod{n}$. In this case, $r = k - j$ is not divisible by $n$ and

$$w^r \neq 1, \quad \text{that is,} \quad w^r - 1 \neq 0.$$

From (4), then, we get that

$$1 + w^r + w^{2r} + \cdots + w^{(n-1)r} = 0,$$

and then (3) yields

$$c_k = \frac{f_k}{n} \cdot 0 = 0.$$

To demonstrate the power of this marvellous result, let us determine the sum $S$ of every third binomial coefficient $\binom{25}{k}$, beginning at the second one:

$$S = \binom{25}{2} + \binom{25}{5} + \binom{25}{8} + \cdots + \binom{25}{23}.$$

This falls out of the binomial series

$$f(x) = (1 + x)^{25} = \binom{25}{0} + \binom{25}{1}x + \binom{25}{2}x^2 + \cdots + \binom{25}{25}x^{25}$$

in the form

$$S(3, 2) = \binom{25}{2}x^2 + \binom{25}{5}x^5 + \cdots + \binom{25}{23}x^{23},$$

from which we obtain the desired sum $S$ by putting $x = 1$.

Our formula gives

$$\begin{aligned} S(3, 2) &= \frac{1}{3} \sum_{t=0}^{2} w^{-2t} f(w^t x) \\ &= \frac{1}{3} [f(x) + w^{-2} f(wx) + w^{-4} f(w^2 x)] \\ &= \frac{1}{3} [(1 + x)^{25} + w^{-2}(1 + wx)^{25} + w^{-4}(1 + w^2 x)^{25}]. \end{aligned}$$

Now, for $n = 3$, we have $w^3 = 1$, and also $1 + w + w^2 = 0$ (in general, the sum of the roots of the equation $y^n - 1 = 0$ is zero, giving $1 + w + w^2 + \cdots + w^{n-1} = 0$). This makes $1 + w = -w^2$ and $1 + w^2 = -w$, and, with $x = 1$, we obtain

$$\begin{aligned} S &= \frac{1}{3} [2^{25} + w^{-2}(1 + w)^{25} + w^{-4}(1 + w^2)^{25}] \\ &= \frac{1}{3} [2^{25} + w^{-2}(-w^2)^{25} + w^{-4}(-w)^{25}] \end{aligned}$$

$$= \frac{1}{3}[2^{25} - w^{48} - w^{21}]$$

$$= \frac{1}{3}(2^{25} - 1 - 1)$$

$$= \frac{1}{3}(2^{25} - 2).$$

This also serves as a nice illustration of the way so-called imaginary numbers can play a major role in performing computations of a completely concrete nature.

CHAPTER **17**

# A PROBLEM OF PAUL ERDÖS

**1.** The great Hungarian mathematician Paul Erdös is no stranger to anyone in the fields of combinatorics, number theory, or geometry. His mathematical output is phenomenal! At present he is the author of more than 1000 first-class papers and hundreds of intriguing problems that have inspired the mathematicians of three generations. On our way to two solutions to one of his recent problems [**1**], we shall encounter several beautiful arguments, beginning with Euler's priceless derivation of the result

$$1+\frac{1}{2^2}+\frac{1}{3^2}+\cdots=\frac{\pi^2}{6}.$$

Professor Erdös has the theory that God has a book containing all the theorems of mathematics with their absolutely most beautiful proofs, and when he wants to express the highest appreciation of a proof he exclaims "This is one from the book!" Erdös does not do this very often, but across the top of the note he sent me, containing our second solution, is the inscription "I think this proof comes straight out of the book!"

The problem concerns a surprising result that was first established in 1974 by Erdös and Stan Benkoski (Pennsylvania State University):

> if $a_1 < a_2 < \cdots < a_k$ is a set of $k$ distinct positive integers *whose* $2^k$ *subsets all have different sums*, then the sum of the reciprocals of the integers must be less than 2:

$$\frac{1}{a_1}+\frac{1}{a_2}+\cdots+\frac{1}{a_k}<2.$$

As is often the case with a good problem, the conclusion seems to be a rather unlikely consequence of the hypothesis.

In our first solution, due to C. Ryavec, we shall make use of Euler's discovery that $\Sigma_{n=1}^{\infty} 1/n^2 = \pi^2/6$. Let us begin, then, with his brilliant 18th century "proof" of this result [**2**]. While his treatment of the infinite processes lacks 20th century rigor, it remains a classic example of first-rate mathematics.

We assume from introductory calculus the well-known series for $\sin x$:

$$\sin x = x - \frac{x^3}{3!} + \frac{x^5}{5!} - \frac{x^7}{7!} + - \cdots.$$

Now, just as a knowledge of the roots 3 and 4 of the equation $x^2 - 7x + 12 = 0$ leads to the factoring $x^2 - 7x + 12 = (x-3)(x-4)$, the roots of $\sin x = 0$, namely, $x = 0, \pm\pi, \pm 2\pi, \ldots, \pm n\pi, \ldots$, lead to a factoring of

$$x - \frac{x^3}{3!} + \frac{x^5}{5!} - + \cdots.$$

The root $x = 0$ provides a factor $x$ and, because the expression is written in increasing powers of $x$, the appropriate form in the present instance is

$$x - \frac{x^3}{3!} + \frac{x^5}{5!} - \cdots$$
$$= x\left(1 - \frac{x}{\pi}\right)\left(1 + \frac{x}{\pi}\right)\left(1 - \frac{x}{2\pi}\right)\left(1 + \frac{x}{2\pi}\right)\cdots.$$

Dividing through by $x$, and noting that $(1 - (x/n\pi))(1 + (x/n\pi)) = 1 - x^2/n^2\pi^2$, we obtain

$$1 - \frac{x^2}{3!} + \frac{x^4}{5!} - \cdots = \left(1 - \frac{x^2}{\pi^2}\right)\left(1 - \frac{x^2}{2^2\pi^2}\right)\left(1 - \frac{x^2}{3^2\pi^2}\right)\cdots.$$

Equating coefficients of $x^2$, we get

$$-\frac{1}{3!} = -\frac{1}{\pi^2} - \frac{1}{2^2\pi^2} - \frac{1}{3^2\pi^2} - \cdots,$$

and, multiplying through by $-\pi^2$, we have the desired

$$\frac{\pi^2}{6} = 1 + \frac{1}{2^2} + \frac{1}{3^2} + \cdots.$$

It happens that the 2 in the final result,

$$\frac{1}{a_1} + \frac{1}{a_2} + \cdots + \frac{1}{a_k} < 2,$$

arises as the quotient of $\pi^2/6$ and $\pi^2/12$. Actually it is the quotient

$$\frac{1 + \frac{1}{2^2} + \frac{1}{3^2} + \frac{1}{4^2} + \cdots}{1 - \frac{1}{2^2} + \frac{1}{3^2} - \frac{1}{4^2} + - \cdots}.$$

From Euler's discovery, however, it is easy to deduce that the series in the denominator is just $\pi^2/12$ (Euler's result implies that the series in the denominator is absolutely convergent, which makes the following manipulations permissible):

$$\begin{aligned} &1 - \frac{1}{2^2} + \frac{1}{3^2} - \frac{1}{4^2} + - \cdots \\ &= 1 + \frac{1}{2^2} + \frac{1}{3^2} + \frac{1}{4^2} + \cdots - 2\left(\frac{1}{2^2} + \frac{1}{4^2} + \frac{1}{6^2} + \cdots\right) \\ &= \frac{\pi^2}{6} - \frac{2}{2^2}\left(1 + \frac{1}{2^2} + \frac{1}{3^2} + \cdots\right) \\ &= \frac{\pi^2}{6} - \frac{1}{2}\cdot\frac{\pi^2}{6} \\ &= \frac{\pi^2}{12}. \end{aligned}$$

With the series for the logarithm function,

$$\log(1 - x) = -x - \frac{x^2}{2} - \frac{x^3}{3} - \cdots, \qquad \text{for } -1 \le x < 1,$$

we are ready to proceed.

**2. Solution 1**

In this first solution, we will not solve the full problem, but will establish only the weaker relation

$$\frac{1}{a_1} + \frac{1}{a_2} + \cdots + \frac{1}{a_k} \le 2.$$

Suppose $0 \le x < 1$. Now, each term in the expansion of the product

$$\begin{aligned} P &= (1 + x^{a_1})(1 + x^{a_2}) \cdots (1 + x^{a_k}) \\ &= \cdots + x^{a_1+0+0+a_4+\cdots+a_k} + \cdots \end{aligned}$$

is a power of $x$ which has for its exponent the sum of one of the $2^k$ subsets of $\{a_1, a_2, \ldots, a_k\}$ (even $1 = x^{0+0+\cdots+0}$).
Since all these subsets have different sums, *no term is duplicated*, implying that every term carries a coefficient of unity:

$$P = 1 + x^{a_1} + x^{a_2} + \cdots + x^{a_1+a_2+\cdots+a_k}.$$

Accordingly, for $x > 0$, $P$ is not as great as the infinite series which contains every power of $x$:

$$P < 1 + x + x^2 + x^3 + \cdots. \tag{1}$$

For $0 \le x < 1$, this series is given by $(1-x)^{-1}$, and since $x = 0$ gives equality in (1), we have for all $x$ under consideration that

$$P = \prod_{i=1}^{k} (1 + x^{a_i}) \le (1 - x)^{-1}. \tag{2}$$

Taking logarithms, this yields

$$\sum_{i=1}^{k} \log(1 + x^{a_i}) \le -\log(1 - x),$$

and the series for the log function gives

$$\sum_{i=1}^{k} \left( x^{a_i} - \frac{x^{2a_i}}{2} + \frac{x^{3a_i}}{3} - + \cdots \right) \le x + \frac{x^2}{2} + \frac{x^3}{3} + \cdots.$$

Since $a_i$ is a positive integer, there is a factor $x$ that is common to every term on both sides of this relation. Dividing through by $x$, then, we obtain an inequality which connects expressions that are defined for $x = 0$:

$$\sum_{i=1}^{k} \frac{\log(1 + x^{a_i})}{x} \leq 1 + \frac{x}{2} + \frac{x^2}{3} + \cdots. \tag{3}$$

Now let us integrate over the interval $[0, t]$, where $0 < t < 1$, to give

$$\sum_{i=1}^{k} \int_0^t \frac{\log(1 + x^{a_i})}{x}\, dx \leq \left[x + \frac{x^2}{2^2} + \frac{x^3}{3^2} + \cdots\right]_0^t$$

$$= t + \frac{t^2}{2^2} + \frac{t^3}{3^2} + \cdots. \tag{4}$$

The left side may be evaluated by standard procedures as follows. In the $i$th integral make the substitution $y = x^{a_i}$; then

$$y = x^{a_i}, \qquad \log y = a_i \cdot \log x, \qquad \frac{dy}{y} = a_i \cdot \frac{dx}{x},$$

and

$$\frac{dx}{x} = \frac{1}{a_i} \cdot \frac{dy}{y}.$$

Now, as $x$ goes from 0 to $t$, $y$ goes from 0 to $t^{a_i}$. Letting $t^{a_i} = s$, we have

$$\int_0^t \frac{\log(1 + x^{a_i})}{x}\, dx = \frac{1}{a_i} \int_0^s \frac{\log(1 + y)}{y}\, dy$$

$$= \frac{1}{a_i} \int_0^s \frac{1}{y} \left(y - \frac{y^2}{2} + \frac{y^3}{3} - \frac{y^4}{4} + - \cdots\right) dy$$

$$= \frac{1}{a_i} \int_0^s \left(1 - \frac{y}{2} + \frac{y^2}{3} - \frac{y^3}{4} + - \cdots\right) dy$$

$$= \frac{1}{a_i} \left[y - \frac{y^2}{2^2} + \frac{y^3}{3^2} - + \cdots\right]_0^s$$

$$= \frac{1}{a_i} \left(s - \frac{s^2}{2^2} + \frac{s^3}{3^2} - + \cdots\right).$$

Thus (4) is

$$\sum_{i=1}^{k} \frac{1}{a_i}\left(s - \frac{s^2}{2^2} + \frac{s^3}{3^2} - + \cdots\right) \le t + \frac{t^2}{2^2} + \frac{t^3}{3^2} + \cdots,$$

where $s = t^{a_i}$. As $t \to 1$ from below, we have $s \to 1$ from below for each positive integer $a_i$, and in the limit we have

$$\sum_{i=1}^{k} \frac{1}{a_i}\left(1 - \frac{1}{2^2} + \frac{1}{3^2} - + \cdots\right) \le 1 + \frac{1}{2^2} + \frac{1}{3^2} + \cdots,$$

$$\frac{\pi^2}{12}\left(\sum_{i=1}^{k} \frac{1}{a_i}\right) \le \frac{\pi^2}{6},$$

giving the desired

$$\frac{1}{a_1} + \frac{1}{a_2} + \cdots + \frac{1}{a_k} \le 2.$$

### 3. Solution 2

Now we come to our coup de grâce, a wonderful solution which uses nothing but simple high-school algebra, by A. Bruen and D. Borwein of the University of Western Ontario, London, Ontario [**1**].

The $2^r$ subsets which can be constructed from the first $r$ of the given integers $a_i$ are among the $2^k$ subsets of the full set $a_1, a_2, \ldots, a_k$, and therefore they must have different sums. Starting at 0 (for the empty set), these $2^r$ different positive integers must attain a value at least as big as $2^r - 1$ (even consecutive integers would go this far), and we have

$$a_1 + a_2 + \cdots + a_r \ge 2^r - 1.$$

Since $2^r - 1 = 1 + 2 + 2^2 + \cdots + 2^{r-1}$, we have, for all $r = 1, 2, 3, \ldots, k$, that

$$a_1 + a_2 + \cdots + a_r \ge 1 + 2 + 2^2 + \cdots + 2^{r-1}.$$

The desired result now follows immediately from the following lemma, with whose easy proof we shall conclude our story:

*if $x_1, x_2, \ldots, x_k$ and $y_1, y_2, \ldots, y_k$ are sets of positive real numbers which satisfy the conditions*

(i) $0 < x_1y_1 < x_2y_2 < \cdots < x_ky_k$,
(ii) $x_1 + x_2 + \cdots + x_r \geq y_1 + y_2 + \cdots + y_r$, *for all* $r = 1, 2, 3, \ldots, k$,

*then*

$$\frac{1}{x_1} + \frac{1}{x_2} + \cdots + \frac{1}{x_k} \leq \frac{1}{y_1} + \frac{1}{y_2} + \cdots + \frac{1}{y_k},$$

*with equality if and only if* $x_r = y_r$ *for all* $r$.

It is not difficult to see that the assignments $x_r = a_r, y_r = 2^{r-1}$ satisfy the conditions of the lemma: since $a_r < a_{r+1}$, we have $x_r < x_{r+1}$; and since $2^{r-1} < 2^r$, we have $y_r < y_{r+1}$; thus we have the required condition (i) $0 < x_1y_1 < \cdots < x_ky_k$; and we have just seen that condition (ii) is satisfied:

$$a_1 + a_2 + \cdots + a_r \geq 1 + 2 + \cdots + 2^{r-1} \quad \text{for all } r,$$

$$\text{i.e.} \quad x_1 + x_2 + \cdots + x_r \geq y_1 + y_2 + \cdots + y_r.$$

Therefore the lemma applies, and we have

$$\begin{aligned} &\frac{1}{a_1} + \frac{1}{a_2} + \cdots + \frac{1}{a_k} \\ &= \frac{1}{x_1} + \frac{1}{x_2} + \cdots + \frac{1}{x_k} \\ &\leq \frac{1}{y_1} + \frac{1}{y_2} + \cdots + \frac{1}{y_k} \\ &= 1 + \frac{1}{2} + \cdots + \frac{1}{2^{k-1}} \\ &= 2 - \frac{1}{2^{k-1}}. \end{aligned}$$

In fact, the bound 2 has even been reduced to $2 - 1/2^{k-1}$.

## 4. Proof of the Lemma

Let us use the notation

$$s_r = (x_1 - y_1) + (x_2 - y_2) + \cdots + (x_r - y_r),$$

$$z_r = \frac{1}{x_r y_r}, \quad \text{for } r = 1, 2, \ldots, k,$$

and

$$z_{k+1} = 0.$$

From the given relation $x_1 + x_2 + \cdots + x_r \geq y_1 + y_2 + \cdots + y_r$, we obtain

$$s_r \geq 0 \tag{5}$$

and from $x_r y_r < x_{r+1} y_{r+1}$, we have $z_r > z_{r+1}$, or

$$z_r - z_{r+1} > 0 \tag{6}$$

(this holds for all $r$, even $r = k$).
Therefore the quantity in question is

$$\left(\frac{1}{y_1} + \frac{1}{y_2} + \cdots + \frac{1}{y_k}\right) - \left(\frac{1}{x_1} + \frac{1}{x_2} + \cdots + \frac{1}{x_k}\right)$$

$$= \left(\frac{1}{y_1} - \frac{1}{x_1}\right) + \left(\frac{1}{y_2} - \frac{1}{x_2}\right) + \cdots + \left(\frac{1}{y_k} - \frac{1}{x_k}\right)$$

$$= \frac{x_1 - y_1}{x_1 y_1} + \frac{x_2 - y_2}{x_2 y_2} + \cdots + \frac{x_k - y_k}{x_k y_k}$$

$$= (x_1 - y_1)z_1 + (x_2 - y_2)z_2 + \cdots + (x_k - y_k)z_k.$$

Now, because $x_1 - y_1 = s_1$ and, in general, $x_r - y_r = s_r - s_{r-1}$, this expression is given by

$$s_1 z_1 + (s_2 - s_1)z_2 + (s_3 - s_2)z_3 + \cdots + (s_k - s_{k-1})z_k$$

$$= s_1(z_1 - z_2) + s_2(z_2 - z_3) + \cdots + s_{k-1}(z_{k-1} - z_k)$$

$$+ s_k(z_k - z_{k+1})$$

By relations (5) and (6), above, each term here is $\geq 0$, making the whole expression $\geq 0$, from which the desired inequality follows.

Because a series of nonnegative terms can only vanish if every term vanishes, we have equality here if and only if each

$$s_r(z_r - z_{r+1}) = 0.$$

Since $z_r - z_{r+1} > 0$, this requires $s_r = 0$ for all $r$, implying

$$s_1 = x_1 - y_1 = 0, \qquad \text{giving } x_1 = y_1;$$

$$s_2 = (x_1 - y_1) + (x_2 - y_2)$$

$$= 0 + (x_2 - y_2) = 0, \quad \text{giving } x_2 = y_2;$$

and so on to $x_r = y_r$ for all $r = 1, 2, \ldots, k$.

Perhaps you might enjoy the following little exercise based on Euler's result:

Prove that

$$\sum_{n=1}^{\infty} \frac{1}{n^4} = 1 + \frac{1}{2^4} + \frac{1}{3^4} + \cdots = \frac{\pi^4}{90}.$$

**References**

1. Canadian Mathematical Bulletin, 17 (1975) 768, Problem P.220.
2. George Polya, Mathematics and Plausible Reasoning, Volume 1, Princeton University Press, Princeton, New Jersey, 1954, pp. 17-21.

CHAPTER **18**

# CAI MAO-CHENG'S SOLUTION TO KATONA'S PROBLEM ON FAMILIES OF SEPARATING SUBSETS

**1.** The 26 letters of the alphabet give rise to $\binom{26}{2} = 325$ pairs of different letters (a, b), (a, c), .... And from this modest pool of 26 letters it it possible to make up a total of $2^{26}$ different subsets, more than 67 million. How many of these subsets do you think it would take to make a family of subsets $(A_1, A_2, \ldots, A_m)$ in which the two letters of each of the 325 pairs can be found to be separated, one in each of two *disjoint* subsets; that is to say, no matter which pair $(x, y)$ might be specified, some two disjoint subsets $A_i$, $A_j$, of the family separate these letters, $x \in A_i$, $y \in A_j$? Of course, it is permissible for $x$ and $y$ also to occur together in some other subset $A_k$; we don't care what else happens so long as some disjoint pair $A_i$, $A_j$ separates them. Let us call such a family of subsets a *separating family*. It might come as a mild surprise to learn that the alphabet can be separated by a family containing as few as 9 subsets (see Figure 1).

In 1973 the Hungarian mathematician G. O. H. Katona posed the general problem of determining, for a set of $n$ elements, the *minimum* number $f(n)$ of subsets in a separating family. This problem was solved in early February, 1982, by the gifted Chinese mathematician Cai Mao-Cheng (Academia Sinica, Peking), during an extended visit to the University of Waterloo. His ingenious solution contains some very pretty mathematics and is a real gem. However, it is quite long and much more complicated than the arguments in the earlier chapters.

| | | |
|---|---|---|
| (*abcdefghi*) | (*jklmnopqr*) | (*stuvwxyz*) |
| (*abcjklstu*) | (*defmnovwx*) | (*ghipqryz*) |
| (*adgjmpsvy*) | (*behknqtwz*) | (*cfilorux*) |

Figure 1

## 2. A Formula for $f(n)$

It turns out that $f(n)$ is the smallest number among three easily-computed values (the notation $\lceil x \rceil$ denotes the so-called *ceiling function*, the smallest integer $\geq x$; that is, if $x$ is not already an integer, round upwards):

$$0 + 3\lceil \log_3 n \rceil, \qquad 2 + 3\left\lceil \log_3 \left(\frac{n}{2}\right) \right\rceil, \qquad 4 + 3\left\lceil \log_3 \left(\frac{n}{4}\right) \right\rceil,$$

that is,

$$f(n) = \min\{2p + 3\lceil \log_3 (n/2^p) \rceil \mid p = 0, 1, 2\}. \tag{1}$$

From the first few powers of 3–3, 9, 27, 81, ... — one can calculate $f(n)$ mentally for small values of $n$. For example,

$$\log_3 26 = 2.\text{----}, \quad \text{making} \quad \lceil \log_3 26 \rceil = 3;$$

similarly

$$\lceil \log_3 13 \rceil = 3, \quad \text{and} \quad \lceil \log_3 (13/2) \rceil = 2;$$

thus

$$\begin{aligned} f(26) &= \min(0 + 3 \cdot 3, 2 + 3 \cdot 3, 4 + 3 \cdot 2) \\ &= \min(9, 11, 10) \\ &= 9. \end{aligned}$$

Since the $n$ subsets, each consisting of a single element, clearly separate the pairs of any set, we always have $f(n) \leq n$. An exhaustive investigation of the cases $n = 2, 3, 4$ reveals that, in these cases, one cannot improve on $f(n) = n$ (this is left to the reader). These values agree with the formula, and to extend this agreement to the trivial case $n = 1$, let us define $f(1) = 0$. The formula is established for $n \geq 4$ by showing that the right side of (1) is both an upper bound and a lower bound on $f(n)$, forcing equality.

We note in passing the obvious fact that increasing $n$ cannot diminish the burden of separating the elements, implying that

$$f(m) \geq f(n) \quad \text{for } m \geq n.$$

### 3. A Fundamental Upper Bound on $f(n)$

Our proof begins with a major lemma.

LEMMA 1. *For* $n \geq 4$,

$$f(n) \leq \min\left\{f(k) + f\left(\left\lceil \frac{n}{k} \right\rceil\right) \mid k = 2, 3, \ldots, n-2\right\}. \quad (2)$$

The idea here is that the operation of distributing the function $f$ over what is essentially a pair of factors of $n$ ($k$ and $\lceil n/k \rceil$) never yields a smaller result. We shall use this later to establish relations such as

$$f(2^p \cdot 3^q) \leq f(2^p) + f(3^q), \quad p, q \text{ nonnegative integers.}$$

This lemma is a very perceptive observation and Cai's proof is quite clever.

Let $S = (x_1, x_2, \ldots, x_n)$ be a set of $n$ elements and let $k$ be any integer in the range $(2, 3, \ldots, n-2)$. Cai proceeds to construct a family of separating subsets for $S$ which has $f(k) + f(\lceil n/k \rceil)$ members as follows. Lay out the elements of $S$ in a rectangular array $W$, putting $k$ elements to a row, as far as they will go:

$$\begin{array}{rllll} & x_1, & x_2, & \ldots, x_k, \\ & x_{k+1}, & x_{k+2}, & \ldots, x_{2k}, \\ W: & x_{2k+1}, & x_{2k+2}, & \ldots, x_{3k}, \\ & \ldots & \ldots & \ldots, \\ & \ldots, & x_{n-1}, & x_n. \end{array}$$

The number of rows in $W$, counting a possibly incomplete last row, is $\lceil n/k \rceil$, which we denote by $m$. Let $Y_i$ denote the $i$th row of $W$, and $Z_i$ the $i$th column.

Armed with the numbers $m$ and $k$, we digress momentarily to introduce two new and independent sets of objects

$$R = (y_1, y_2, \ldots, y_m), \quad \text{and} \quad T = (z_1, z_2, \ldots, z_k).$$

Because a separating family must exist for every set $X = (a_1, a_2, \ldots, a_r)$ (for example, the set of singletons), there must be one having the fewest subsets, that is, $f(r)$ subsets. Accordingly, let $D$ and $E$ be separating families for $R$ and $T$, respectively, which contain the minimum number of subsets $f(m)$ and $f(k)$:

$$D = \{D_1, D_2, \ldots, D_{f(m)}\} \quad \text{for} \quad (y_1, y_2, \ldots, y_m),$$

and

$$E = \{E_1, E_2, \ldots, E_{f(k)}\} \quad \text{for} \quad (z_1, z_2, \ldots, z_k).$$

(Recall that this means that, for any pair $(y_s, y_t)$ of $R$, there is a pair of disjoint subsets $(D_u, D_v)$ of $D$ such that $y_s \in D_u$, $y_t \in D_v$; similarly for $T$. Recall also that the $m$ rows of $W$ are $Y_1, Y_2, \ldots, Y_m$, and the $k$ columns are $Z_1, Z_2, \ldots Z_k$). The procedure is to replace the elements $y_j$ in the subsets $D_i$ by the rows $Y_j$: if $D_i = (y_a, y_b, \ldots)$, we construct $B_i = (Y_a, Y_b, \ldots)$, making $B_i$ a subset of $S = (x_1, x_2, \ldots, x_n)$. Since no two $x$'s in $W$ are the same, no two $x$'s in $B_i$ are the same. In this way the $f(m)$ subsets $D_1, D_2, \ldots, D_{f(m)}$ give rise to $f(m)$ subsets $B_1$, $B_2, \ldots B_{f(m)}$. Similarly, by changing all the $z$'s in the $E$'s to $Z$'s, the sets $E_1, E_2, \ldots E_{f(k)}$ yield a collection of subsets $C_1, C_2, \ldots, C_{f(k)}$ of $S$.

It is easy to show that the family

$$B_1, B_2, \ldots, B_{f(m)}, C_1, C_2, \ldots, C_{f(k)}$$

is a separating family for $S$. To this end, let $x_i, x_j$ be any two different elements of $S$. These elements occur somewhere in the array $W$, and if they don't lie in different rows, they must surely lie in different columns (they cannot both be in the same row *and* in the same column).

(i) *Different Rows*. Suppose $x_i \in Y_s$, $x_j \in Y_t$, where $s \neq t$. With the subscripts $s$ and $t$ in hand, we switch attention to the set $R = (y_1, y_2, \ldots, y_m)$ and its separating family $D_1, D_2, \ldots, D_{f(m)}$. Corresponding to the pair $(y_s, y_t)$ there exists a pair of disjoint subsets $D_u$, $D_v$ which separates them, say, $y_s \in D_u$ and $y_t \in D_v$. In converting from the $D$'s to the $B$'s, then, because $y_s \in D_u$, the row $Y_s$ is put into $B_u$. Similarly $Y_t$ belongs to $B_v$. But $D_u$ and $D_v$ are disjoint, implying that none of the rows $Y_j$ that go into $B_u$ also go into $B_v$. Thus $B_u$ and $B_v$ must be disjoint subsets of $S$. Since $x_i \in Y_s$, and $Y_s \subseteq B_u$, we have $x_i \in B_u$; similarly $x_j \in B_v$, and the pair $B_u$, $B_v$ are disjoint subsets of $S$ that separate $x_i$ *and* $x_j$.

(ii) *Different Columns*. This is an exactly parallel case.

Therefore, the family $B_1, \ldots, B_{f(m)}, C_1, \ldots, C_{f(k)}$, containing $f(m) + f(k)$ subsets, does indeed constitute a separating family for $S$. Consequently $f(n)$, the minimum number of subsets in such a family, cannot exceed $f(m) + f(k)$. Since $m = \lceil n/k \rceil$, and $k$ is an arbitrary integer in the range $(2, 3, \ldots, n-2)$, we obtain the desired result

$$f(n) \leq \min\left\{f(k) + f\left(\left\lceil \frac{n}{k} \right\rceil\right) \mid k = 2, 3, \ldots, n-2\right\}.$$

Let us use this immediately to prove the following corollary, which we will need later.

COROLLARY. *For* $n \geq 6$, $f(n) \leq n - 1$.

It is not difficult to see that, for $n \geq 6$, dropping from the value $n$ to the integer $\lceil n/2 \rceil$ involves a loss of at least 3 units, that is,

$$\left\lceil \frac{n}{2} \right\rceil \leq n - 3.$$

By lemma 1, then, we have

$$\begin{aligned} f(n) &\leq \min\left\{f(k) + f\left(\left\lceil \frac{n}{k} \right\rceil\right) \mid k = 2, 3, \ldots, n-2\right\}. \\ &\leq f(2) + f\left(\left\lceil \frac{n}{2} \right\rceil\right) \quad \text{(a minimum is } \leq \text{ any particular value)} \\ &\leq 2 + \left\lceil \frac{n}{2} \right\rceil \qquad (f(r) \text{ is always } \leq r) \\ &\leq 2 + (n-3) \\ &= n - 1. \end{aligned}$$

## 4. A Fundamental Lower Bound on $f(n)$

Next we establish a second major lemma.

LEMMA 2. *For* $n \geq 4$,

$$f(n) \geq \min\left\{k + f\left(\left\lceil \frac{n}{k} \right\rceil\right) \mid k = 2, 3, \ldots, n-2\right\}. \tag{3}$$

Suppose $S = (x_1, x_2, \ldots, x_n)$ and that $A = (A_1, A_2, \ldots, A_{f(n)})$ is a separating family for $S$.

Now the subsets $A_i$ generally contain varying numbers of $x$'s. Let $b$ denote the greatest number of $x$'s in any of the $A_i$. Clearly either $b = 1$ or $b \geq 2$.

(i) *Suppose* $b = 1$. In this case, $A$ must be the trivial family of singletons, and $f(n) = n$. In establishing (3), we observe that if $f(n)$ is as great as any particular value of $k + f(\lceil n/k \rceil)$, it is certainly as big as the minimum such value.

For $n \geq 4$, it is easy to see that the number $\lceil n/2 \rceil$ cannot exceed $n - 2$. Now, for $k = 2$, the particular value of $k + f(\lceil n/k \rceil)$ is

$$\begin{aligned} 2 + f\left(\left\lceil \frac{n}{2} \right\rceil\right) &\leq 2 + \left\lceil \frac{n}{2} \right\rceil \\ &\leq 2 + (n - 2) \\ &= n \\ &= f(n). \end{aligned}$$

Thus (3) is valid if $b = 1$.

(ii) *Suppose* $b \geq 2$. Cai disposes of this case in grand style with a brilliant way of dividing the subsets of $A$ into two classes. First select any subset $\bar{A}$ containing the maximum number $b$ of $x$'s. Then simply divide the subsets $A_i$ into the ones which are disjoint to $\bar{A}$ and those which are not:

$$\begin{aligned} \text{class } 1 &= (A_i \mid A_i \cap \bar{A} = \emptyset); \\ \text{class } 2 &= (A_i \mid A_i \cap \bar{A} \neq \emptyset). \end{aligned}$$

We observe that $\bar{A}$ itself belongs to class 2.

Now the subscripts $i$ of the $A_i$ are just identifying labels. Let the subsets be renamed so that the subsets in class 1, in any order, are $A_1, A_2, \ldots, A_{k-1}$, the set $\bar{A}$ is $A_k$, and the rest of the subsets (namely, class 2), in any order, are $A_{k+1}, A_{k+2}, \ldots, A_{f(n)}$.

Now let all the $x$'s in the $A$'s of class 1 be pooled together into a subset $S_1$ of $S$; and let $S_2$ be the complementary subset:

$$\begin{aligned} S_1 &= (x_j \mid x_j \in A_i, \qquad i = 1, 2, \ldots, k - 1), \\ S_2 &= S - S_1. \end{aligned}$$

We should observe that while $S_1$ consists precisely of all the $x$'s in the subsets of class 1, $S_2$ does not necessarily contain all the $x$'s in the subsets of class 2. Consider $x_4$ in the illustration in Figure 2. It could easily happen that a subset $A_i$ of class 1, while certainly being disjoint from $A_k$, shares an $x_r$ (e.g., $x_4$) with some other subset $A_j$ of class 2; being in class 1, this $x_r$ goes into $S_1$, not $S_2$, even though it is present in a subset of class 2.

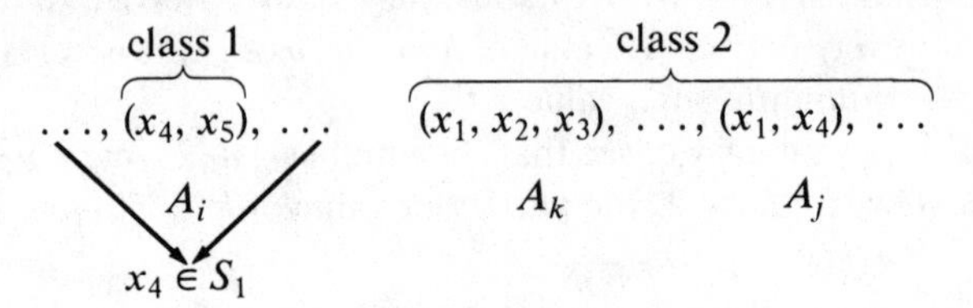


Figure 2

Now let's do some simple counting. Since $S_1$ and $S_2$ are complementary subsets, the total number of elements between them is just $n$:

$$|S_1| + |S_2| = |S| = n.$$

Even though $S_2$ might not contain all the $x$'s in every subset of class 2, it does receive the entire contents of $A_k$, itself. Thus $S_2$ must contain at least the $b$ elements of $A_k$, and we have

$$|S_2| \geq b.$$

Because $S_1$ gains its elements from $k - 1$ subsets $A_i$, each of which contains not more than $b$ elements, it could not possess altogether more than $(k - 1)b$ members, and we obtain

$$|S_1| \leq (k - 1)b \leq (k - 1)|S_2|.$$

Adding $|S_2|$ to each side gives

$$|S_1| + |S_2| \leq k|S_2|,$$

$$n \leq k|S_2|,$$

and

$$|S_2| \geq \frac{n}{k}.$$

Since $|S_2|$ is an integer, this is equivalent to

$$|S_2| \geq \left\lceil \frac{n}{k} \right\rceil.$$

We have seen that $S_2$ might not get all of the $x$'s in the subsets of class 2. However, whatever $x$'s it does get must certainly come from these subsets (it does get all of $A_k$, at least). We focus now on the parts of these subsets, other than $A_k$ itself, which *do* go into the composition of $S_2$. These are the intersections

$$S_2 \cap A_{k+1}, \quad S_2 \cap A_{k+2}, \ldots, \quad S_2 \cap A_{f(n)}.$$

In fact, let us show that these subsets constitute a separating family for $S_2$.

Accordingly, let $x_i$, $x_j$ be any pair of elements in $S_2$. Because the family

$$A_1, A_2, \ldots, A_{f(n)}$$

is a separating family for the whole set $S$, there must exist a pair of disjoint subsets $A_s$, $A_t$ such that

$$x_i \in A_s \quad \text{and} \quad x_j \in A_t.$$

Because $x_i$ and $x_j$ come from $S_2$, they cannot belong to any of the subsets $A_1, A_2, \ldots, A_{k-1}$ (all of whose elements go into $S_1$). Therefore $A_s$ and $A_t$ must be among $(A_k, A_{k+1}, \ldots, A_{f(n)})$. But $A_s$ and $A_t$ are disjoint. Thus neither of them could be $A_k$ for, by definition, none of $A_{k+1}, A_{k+2}, \ldots, A_{f(n)}$ is disjoint from $A_k$. Hence

$$A_s, A_t \in (A_{k+1}, A_{k+2}, \ldots, A_{f(n)}).$$

Since $x_i$ belongs to both $A_s$ and $S_2$, it must lie in the intersection $S_2 \cap A_s$; similarly, $x_j \in S_2 \cap A_t$. Since $A_s$ and $A_t$ are disjoint, their intersections with $S_2$ must also be disjoint, and we have found disjoint subsets among the family

$$S_2 \cap A_{k+1}, \ldots, S_2 \cap A_{f(n)}$$

that separate $x_i$ and $x_j$. Thus this family is indeed a separating family for $S_2$.

Because this particular family contains $f(n) - k$ subsets, the minimum such number cannot exceed this value, and we have

$$f(|S_2|) \le f(n) - k.$$

Using $|S_2| \ge \lceil n/k \rceil$, which we deduced earlier, we get

$$\begin{aligned} f(n) - k &\ge f(|S_2|) \\ &\ge f\left(\left\lceil \frac{n}{k} \right\rceil\right) \qquad (f(m) \ge f(n) \text{ for } m \ge n), \end{aligned}$$

and

$$f(n) \ge k + f\left(\left\lceil \frac{n}{k} \right\rceil\right). \tag{4}$$

Now if $f(n)$ is at least as big as any one of the values $k + f(\lceil n/k \rceil)$, for $k = 2, 3, \ldots, n - 2$, it will certainly be as great as the minimum such value. We have shown in (4) that

$$f(n) \ge k + f\left(\left\lceil \frac{n}{k} \right\rceil\right),$$

where $k$ denotes the position of the special subset $\bar{A}$ when the subsets of the separating family $A$ are renumbered. It remains, then, to show that such a renumbering must always place the number $k$ among the integers $2, 3, \ldots, n - 2$. But this is easy to do.

We have already established the following 3 relations:

$$\begin{aligned} &\text{(i) } f(n) \le n, \\ &\text{(ii) } |S_2| \ge b \ge 2, \\ &\text{(iii) } f(n) - k \ge f(|S_2|), \end{aligned}$$

where $k$ is the value that is generated by the renumbering operation. For such $k$, then, we have

$$n - k \ge f(n) - k \ge f(|S_2|) \ge f(2) = 2,$$

yielding

$$n - k \ge 2 \quad \text{and} \quad k \le n - 2.$$

On the other hand, if $k$ were to be as small as 1, then $A_k$ would be $A_1$ and, because $A_k$ is *not disjoint* from the subsets $A_{k+1}, A_{k+2}, \ldots,$

$A_{f(n)}$, the subset $A_k$ would be completely useless to the family in affecting the separation of the elements of $S$; the remaining $f(n) - 1$ subsets would have to do all the separating themselves, and would thus constitute a separating family containing fewer than $f(n)$ subsets, which is impossible by definition. Thus the $k$ in (4) does lie in the range $(2, 3, \ldots, n-2)$, and (3) is established.

## 5. The Final Step

Now we can prove the main result.

THEOREM. *For* $n \geq 1$,

$$f(n) = \min\{2p + 3\lceil \log_3(n/2^p)\rceil \mid p = 0, 1, 2\}. \tag{5}$$

(i) We have seen that this is valid for $n \leq 4$ and, in considering the cases $n \geq 5$, the lemmas come into play. Let $k_0$ denote the value of $k$ in the range $(2, 3, \ldots, n-2)$ which makes $k + f(\lceil n/k \rceil)$ a minimum. Then, using the lemmas, we have

$$k_0 + f\left(\left\lceil \frac{n}{k_0} \right\rceil\right) \leq f(n) \leq \min\left\{f(k) + f\left(\left\lceil \frac{n}{k} \right\rceil\right) \mid k = 2, 3, \ldots, n-2\right\}$$

$$\leq f(k_0) + f\left(\left\lceil \frac{n}{k_0} \right\rceil\right) \quad \text{(a minimum} \leq \text{a particular value)}$$

$$\leq k_0 + f\left(\left\lceil \frac{n}{k_0} \right\rceil\right) \quad (f(r) \leq r).$$

Since the first and last of these quantities are the same, we must have equality throughout and, for $n \geq 5$, we conclude that

$$f(n) = k_0 + f\left(\left\lceil \frac{n}{k_0} \right\rceil\right).$$

(ii) Now let's consider the desired equation (5). To simplify the expressions, let

$$\lceil \log_3 (n/2^p) \rceil = q.$$

In this case, we have

$$3^q = 3^{\lceil \log_3(n/2^p)\rceil} \geq 3^{\log_3(n/2^p)} = n/2^p,$$

giving $2^p \cdot 3^q \geq n$. Since $p = 0, 1$, or $2$, the biggest that $2^p$ can be is 4, and since $n > 4$, this forces $3^q > 1$, making the integer $q \geq 1$. Thus, for $n > 4$, the only value less than 6 that $2^p \cdot 3^q$ is capable of taking is $2^0 \cdot 3^1 = 3$. Because $2^p \cdot 3^q \geq n > 4$, then, we conclude that $2^p \cdot 3^q$ is always at least 6. We will use this little fact shortly.

We observe that equation (5) is the claim

$$f(n) = \min\{2p + 3q \mid p = 0, 1, 2\}.$$

As noted earlier, this will be established by showing that the right side is both an upper bound and a lower bound on $f(n)$.

(a) $f(n) \leq \min\{2p + 3q\}$. Because $n \leq 2^p \cdot 3^q$, it is clear that $f(n) \leq f(2^p \cdot 3^q)$. Now our plan is to apply lemma 1 to $f(2^p \cdot 3^q)$ to extend this to

$$f(n) \leq f(2^p \cdot 3^q) \leq f(2^p) + f(3^q), \tag{6}$$

and then to show that $f(2^p) \leq 2p$ and $f(3^q) \leq 3q$.

In applying lemma 1 to $f(2^p \cdot 3^q)$, the value of $k$ that is of interest, of course, is

$$k = 2^p, \quad \text{making} \quad \left\lceil \frac{(2^p \cdot 3^q)}{2^p} \right\rceil = 3^q$$

(again we use the idea that $f \leq$ minimum implies $f \leq$ any particular value). Because $p = 0, 1$, or $2$, however, the value of $2^p$ does not necessarily lie in the range of $k$, namely $(2, 3, \ldots, 2^p \cdot 3^q - 2)$. Since $2^p \cdot 3^q$ is always at least 6 for the $n$ under consideration, this range does always accommodate both $2^1$ and $2^2$ (establishing the case of $p = 1, 2$ by lemma 1). Still, we need to consider separately the case of $p = 0$. But this is a trivial matter, for in this case (6) merely makes the obvious claim

$$f(3^q) \leq f(1) + f(3^q).$$

Therefore, relation (6) is indeed universally valid.

We complete this section by showing that

$$\text{(i)} \quad f(2^p) \leq 2p \quad \text{and} \quad \text{(ii)} \quad f(3^q) \leq 3q.$$

There are only three cases of (i), each of which clearly verifies the claim:

$$f(1) = 0 \leq 2(0); \qquad f(2) = 2 \leq 2(1); \qquad f(4) = 4 \leq 2(2).$$

And (ii) is an easy induction. For $q = 1$ we have

$$f(3^q) = f(3) = 3 \leq 3(1),$$

and if $f(3^{q-1}) \leq 3(q-1)$ for some $q \geq 2$ (making $3^q \geq 9$), then

$$\begin{aligned} f(3^q) &\leq \min\left\{ \mathrm{f}(k) + f\left( \left\lceil \frac{3^q}{k} \right\rceil \right) \mid k = 2, 3, \ldots, 3^q - 2 \right\} \\ &\leq f(3) + f(3^{q-1}) \qquad \text{(the value for } k = 3) \\ &\leq 3 + 3(q-1) \\ &= 3q, \qquad \text{as required.} \end{aligned}$$

(b) It remains to establish the lower bound (7) on $f(n)$.

$$f(n) \geq \min\{2p + 3q\}. \tag{7}$$

If $f(n)$ is as great as any of the values taken by $(2p + 3q)$, then it must be as great as the minimum such value. Accordingly, let us show that at least one of the three values taken by $(2p + 3q)$ does not exceed $f(n)$. We proceed by induction.

For $n = 4$, the values of $(2p + 3q)$ are 6, 5, and 4; since $f(4) = 4$, relation (7) holds. Similarly, it is easy to check that (7) holds for all $n \leq 4$. Suppose, then, that $n > 4$ and that (7) holds for all values $\leq n - 1$. Now consider (7) for the value $n$.

Our first step in this final stage revealed the result that the value $k_0$ which yields the minimum value of $f(k) + f(\lceil n/k \rceil)$ is subject to the relation

$$k_0 + f\left( \left\lceil \frac{n}{k_0} \right\rceil \right) \leq f(k_0) + f\left( \left\lceil \frac{n}{k_0} \right\rceil \right),$$

*giving* $k_0 \leq f(k_0)$. But clearly

$$f(k_0) \leq k_0 \qquad (f(r) \text{ is always } \leq r),$$

and it follows that

$$f(k_0) = k_0.$$

Now, by the corollary to lemma 1, we know that, for $n \geq 6$, the value of $f(n)$ does not exceed $n - 1$. Consequently, it must be that

$$k_0 \leq 5.$$

Since the range of $k$ begins at 2, then, we have

$$k_0 = 2, 3, 4, \text{ or } 5.$$

In each case we shall see that (7) is valid, completing the proof.

(i) $k_0 = 2$. From the earlier result $f(n) = k_0 + f(\lceil n/k_0 \rceil)$, we have, with $k_0 = 2$, that

$$f(n) = 2 + f\left(\left\lceil \frac{n}{2} \right\rceil\right).$$

Now the three values on the right side of (7) are

$$\begin{aligned} v_0 &= 3 \lceil \log_3(n/1) \rceil, \\ v_1 &= 2 + 3 \lceil \log_3(n/2) \rceil, \\ v_2 &= 4 + 3 \lceil \log_3(n/4) \rceil. \end{aligned}$$

We shall see that, in all cases, $f(n)$ is at least as big as one of these numbers.

Since $\lceil n/2 \rceil < n$, the induction hypothesis gives

$$f\left(\left\lceil \frac{n}{2} \right\rceil\right) \geq \min\left\{2p + 3\left\lceil \log_3\left(\left\lceil \frac{n}{2} \right\rceil \Big/ 2^p\right)\right\rceil \Big| \, p = 0, 1, 2\right\}.$$

Now suppose that $p_0$ is the value of $p$ that yields the minimum value here. In this case we have

$$\begin{aligned} f(n) &= 2 + f\left(\left\lceil \frac{n}{2} \right\rceil\right) \quad \text{(just shown above)} \\ &\geq 2 + \left[2p_0 + 3\left\lceil \log_3\left(\left\lceil \frac{n}{2} \right\rceil \Big/ 2^{p_0}\right)\right\rceil\right] \\ &\geq 2(p_0 + 1) + 3 \lceil \log_3(n/2^{p_0+1}) \rceil \end{aligned}$$

(dropping $\lceil \;\; \rceil$ cannot increase a value).

The right side of this result is just $v_1$ if $p_0 = 0$, and $v_2$ if $p_0 = 1$, giving the desired conclusion in these cases. Otherwise $p_0$ can only be 2, in which case this relation becomes

$$\begin{aligned} f(n) &\geq 2(3) + 3\left\lceil \log_3\left(\frac{n}{8}\right)\right\rceil \\ &= 3\left\lceil 2 + \log_3\left(\frac{n}{8}\right)\right\rceil \qquad \text{(clearly, if } r \text{ is an integer, } \lceil r + x\rceil = r + \lceil x\rceil\text{)} \\ &= 3\left\lceil \log_3 9 + \log_3\left(\frac{n}{8}\right)\right\rceil \\ &= 3\left\lceil \log_3\left(\frac{9n}{8}\right)\right\rceil \\ &\geq 3\lceil \log_3 n\rceil \\ &= v_0, \end{aligned}$$

and the conclusion follows.

(ii) $k_0 = 3$. Proceeding similarly, we have

$$\begin{aligned} f(n) &= 3 + f\left(\left\lceil\frac{n}{3}\right\rceil\right) \\ &\geq 3 + \min\left\{2p + 3\left\lceil \log_3\left(\left\lceil\frac{n}{3}\right\rceil \Big/ 2^p\right)\right\rceil \,\Big|\, p = 0, 1, 2\right\} \\ &\geq \min\left\{3 + 2p + 3\left\lceil \log_3\left(\frac{n}{3}\Big/ 2^p\right)\right\rceil \,\Big|\, p = 0, 1, 2\right\} \\ &= \min\left\{2p + 3\left\lceil 1 + \log_3\left(\frac{n}{3}\Big/ 2^p\right)\right\rceil \,\Big|\, p = 0, 1, 2\right\} \\ &= \min\{2p + 3\lceil \log_3(n/2^p)\rceil \,|\, p = 0, 1, 2\} \end{aligned}$$

(since $1 = \log_3 3$), establishing (7) directly.

(iii) $k_0 = 4$. Again, we have similarly, that

$$
\begin{aligned}
f(n) &= 4 + f\left(\left\lceil \frac{n}{4} \right\rceil\right) \\
&\geq 4 + \min\left\{2p + 3\left\lceil \log_3\left(\left\lceil \frac{n}{4} \right\rceil \Big/ 2^p\right)\right\rceil \,\Big|\, p = 0, 1, 2\right\} \\
&= \min\left\{2(p+2) + 3\left\lceil \log_3\left(\left\lceil \frac{n}{4} \right\rceil \Big/ 2^p\right)\right\rceil \,\Big|\, p = 0, 1, 2\right\} \\
&= 2(p_0 + 2) + 3\left\lceil \log_3\left(\left\lceil \frac{n}{4} \right\rceil \Big/ 2^{p_0}\right)\right\rceil
\end{aligned}
$$

(where $p_0$ is the value of $p$ which yields the minimum)

$$\geq 2(p_0 + 2) + 3\lceil \log_3(n/2^{p_0+2}) \rceil .$$

If $p_0 = 0$, the right side here is $v_2$, and (7) is satisfied.
Otherwise $p_0 = 1$ or $2$, and we proceed

$$
\begin{aligned}
f(n) &\geq 2(p_0 + 2) + 3\lceil \log_3(n/2^{p_0+2}) \rceil \\
&= 2(p_0 - 1) + 3\lceil 2 + \log_3(n/2^{p_0+2}) \rceil \\
&\qquad \text{(subtracting and adding 6)} \\
&= 2(p_0 - 1) + 3\lceil \log_3(9n/2^{p_0+2}) \rceil \\
&\geq 2(p_0 - 1) + 3\lceil \log_3(n/2^{p_0-1}) \rceil \qquad \text{(since } 9/8 > 1\text{)}.
\end{aligned}
$$

Since the right side is $v_0$ for $p_0 = 1$, and $v_1$ for $p_0 = 2$, (7) is always valid.

(iv) $k_0 = 5$. Finally we come to

$$f(n) = 5 + f\left(\left\lceil \frac{n}{5} \right\rceil\right).$$

By lemma 1, we have

$$
\begin{aligned}
f\left(2\left\lceil \frac{n}{5} \right\rceil\right) &\leq \min\left\{ f(k) + f\left(\left\lceil \frac{2\left\lceil \frac{n}{5} \right\rceil}{k} \right\rceil\right) \,\Big|\, k = 2, \ldots, 2\left\lceil \frac{n}{5} \right\rceil - 2 \right\} \\
&\leq f(2) + f\left(\left\lceil \frac{n}{5} \right\rceil\right) \qquad \text{(with } k = 2\text{)},
\end{aligned}
$$

from which we obtain that

$$f\left(\left\lceil \frac{n}{5} \right\rceil\right) \geq f\left(2\left\lceil \frac{n}{5} \right\rceil\right) - f(2) = f\left(2\left\lceil \frac{n}{5} \right\rceil\right) - 2.$$

Therefore

$$f(n) = 5 + f\left(\left\lceil \frac{n}{5} \right\rceil\right)$$

yields

$$\begin{aligned} f(n) &\geq 5 + f\left(2\left\lceil \frac{n}{5} \right\rceil\right) - 2 \\ &= 3 + f\left(2\left\lceil \frac{n}{5} \right\rceil\right). \end{aligned}$$

But, for $n > 4$, it is not difficult to show that $2\lceil n/5 \rceil \geq \lceil n/3 \rceil$ (this is left to the reader), giving

$$f(n) \geq 3 + f\left(\left\lceil \frac{n}{3} \right\rceil\right).$$

Since

$$f(n) = \min\left\{k + f\left(\left\lceil \frac{n}{k} \right\rceil\right)\right\},$$

the particular value $3 + f(\lceil n/3 \rceil)$ cannot be less than $f(n)$ (since $n \geq 5$, we have $n - 2 \geq 3$, placing 3 in the range of $k$). Accordingly,

$$f(n) = 3 + f\left(\left\lceil \frac{n}{3} \right\rceil\right),$$

and we conclude that the minimum value of $k + f(\lceil n/k \rceil)$ is also given by $k = 3$. That is to say, $k_0 = 5$ is a joint case with $k_0 = 3$. Since we already have seen that (7) is satisfied when $k_0 = 3$, the proof is complete!

## Reference

1. G. O. H. Katona, Combinatorial search problem, in Jagdish N. Srivastava et al, eds, A Survey of Combinatorial Theory, North-Holland, Amsterdam, 1973, pp. 285–308.

# SOLUTIONS TO SELECTED EXERCISES

## Two Problems on Generating Functions

1. The generating function for the partitions in question are

(i) no even part repeated: $\Pi_{k\geq 1}(1-x^{2k-1})^{-1}(1+x^{2k})$;

(ii) no part occurs more than 3 times: $\Pi_{k\geq 1}(1+x^k+x^{2k}+x^{3k})$;

(iii) no part is divisible by 4: $\Pi_{k\geq 1}(1-x^{4k})/(1-x^k)$.

Easy manipulations show that (i) and (ii) are the same as (iii).

2. The generating function for the number of partitions in which no part occurs more than $d$ times is

$$f(x)=\prod_{k\geq 1}(1+x^k+x^{2k}+\cdots+x^{dk})=\prod_{k\geq 1}\frac{1-x^{(d+1)k}}{1-x^k}.$$

The generating function for the number of partitions in which no term is a multiple of $d+1$ is clearly the same as this latter expression for $f(x)$.

3. The generating function for the number of partitions in which every part occurs either 2, 3, or 5 times is

$$\prod_{k\geq 1}(1+x^{2k}+x^{3k}+x^{5k})=\prod_{k\geq 1}(1+x^{2k})(1+x^{3k})$$

$$=\prod_{k\geq 1}\frac{1-x^{4k}}{1-x^{2k}}\cdot\frac{1-x^{6k}}{1-x^{3k}}.$$

In the denominator, the factor $1 - x^{2k}$ provides a term $1 - x^i$ for $i =$ 2, 4, 6, 8, 10, 12, 14, ..., and $1 - x^{3k}$ provides a term $1 - x^i$ for $i =$ 3, 6, 9, 12, 15, .... The $1 - x^{4k}$ in the numerator cancels terms $1 - x^i$ for $i = 4, 8, 12, 16, \ldots$, and the $1 - x^{6k}$ cancels $1 - x^i$ for $i = 6$, 12, 18, .... Therefore the terms remaining in the denominator are $1 - x^i$ for $i = 2, 6, 10, 14, \ldots$, and $i = 3, 9, 15, \ldots$, that is, for $i \equiv$ 2, 3, 6, 9, 10 (mod 12). Hence the conclusion.

4. The generating function for the partitions in which no part occurs exactly once is

$$\prod_{k\geq 1} (1 + x^{2k} + x^{3k} + \cdots) = \prod_{k\geq 1} [1 + x^{2k}(1 + x^k + x^{2k} + \cdots)]$$

$$= \prod_{k\geq 1} [1 + x^{2k}(1 - x^k)^{-1}] = \prod_{k\geq 1} [(1 - x^k)^{-1}(1 - x^k + x^{2k})]$$

$$= \prod_{k\geq 1} (1 - x^k)^{-1} \cdot \frac{1 + x^{3k}}{1 + x^k} = \prod_{k\geq 1} \frac{1 + x^{3k}}{1 - x^{2k}} = \prod_{k\geq 1} \frac{1 + x^{6k}}{(1 - x^{2k})(1 - x^{3k})}.$$

In the denominator, the factor $1 - x^{2k}$ provides terms $1 - x^i$ for $i =$ 2, 4, 6, 8, ..., and $1 - x^{3k}$ provides terms $1 - x^i$ for $i = 3, 6, 9, \ldots$. The $1 - x^{6k}$ in the numerator cancels $1 - x^i$ for $i = 6, 12, 18, \ldots$, leaving terms for $i = 2, 3, 4, 6, 8, 9, 10, 12, \ldots$, that is, those for $i \equiv$ 0, 2, 3, 4 (mod 6). Hence the conclusion.

**A Second Look at the Fibonacci and Lucas Numbers**

**Posed in section on divisibility.**

For odd $f_n$ we have $(17, f_n) = (34, f_n) = (f_9, f_n) = f_{(9,n)} = f_1, f_3$, or $f_9$, that is, 1, 2, or 34. Since $(17, f_n)$ must be odd, then $(17, f_n) =$ 1, and $17 \nmid f_n$.

**Posed in the set of exercises at the end.**

2. Since $f_n = f_{n+2} - f_{n+1}$, we have

$$S = \sum_{n=1}^{\infty} \frac{f_n}{f_{n+1}f_{n+2}} = \sum_{n=1}^{\infty} \frac{f_{n+2} - f_{n+1}}{f_{n+1}f_{n+2}}$$

$$= \sum_{n=1}^{\infty} \left[ \frac{1}{f_{n+1}} - \frac{1}{f_{n+2}} \right] \quad \text{(in which both terms have the \textit{same form})}$$

$$= \left(\frac{1}{f_2} - \frac{1}{f_3}\right) + \left(\frac{1}{f_3} - \frac{1}{f_4}\right) + \left(\frac{1}{f_4} - \frac{1}{f_5}\right) + \cdots,$$

having partial sums $S_n = (1/f_2) - (1/f_{n+2})$. Since $1/f_{n+2} \to 0$ as $n \to \infty$, we have $S = \lim_{n\to\infty} S_n = 1/f_2 = 1$.

3. We know that $L_m | f_n$ if and only if $m$ divides into $n$ an even number of times. Since $L_4 = 7$, we have $7 | f_n$ if and only if 4 divides into $n$ an even number of times, which requires $n$ to be even. If $n$ is odd, we never have $7 | f_n$.

5. We have

$$f_{2n}(L_{2n}^2 - 1) = \left(\frac{\alpha^{2n} - \beta^{2n}}{\alpha - \beta}\right) [(\alpha^{2n} + \beta^{2n})^2 - 1]$$

$$= \frac{1}{\alpha - \beta} (\alpha^{2n} - \beta^{2n})(\alpha^{4n} + 2\alpha^{2n}\beta^{2n} + \beta^{4n} - 1)$$

$$= \frac{1}{\alpha - \beta} (\alpha^{2n} - \beta^{2n})(\alpha^{4n} + 1 + \beta^{4n}) \qquad (\text{since } \alpha\beta = -1)$$

$$= \frac{1}{\alpha - \beta} (\alpha^{6n} + \alpha^{2n} + \alpha^{2n}\beta^{4n} - \beta^{2n}\alpha^{4n} - \beta^{2n} - \beta^{6n})$$

$$= \frac{1}{\alpha - \beta} (\alpha^{6n} + \alpha^{2n} + \beta^{2n} - \alpha^{2n} - \beta^{2n} - \beta^{6n})$$

(since $\alpha\beta = -1$, then $\alpha^{2n}\beta^{4n} = \beta^{2n}$, for example)

$$= \frac{1}{\alpha - \beta} (\alpha^{6n} - \beta^{6n}) = f_{6n}.$$

This gives $f_{2n} + f_{6n} = f_{2n}L_{2n}^2$, and we have

$$f_{4n}^2 + 8f_{2n}(f_{2n} + f_{6n}) = f_{4n}^2 + 8f_{2n} \cdot f_{2n}L_{2n}^2$$

$$= f_{4n}^2 + 8f_{2n}^2L_{2n}^2 = f_{4n}^2 + 8f_{4n}^2 \qquad (\text{since } f_{2n}L_{2n} = f_{4n})$$

$$= 9f_{4n}^2 = (3f_{4n})^2, \qquad \text{a perfect square.}$$

6. (ii). We have

$$L = \lim_{n\to\infty} \sqrt[n]{f_n} = \lim_{n\to\infty} \left(\frac{\alpha^n - \beta^n}{\alpha - \beta}\right)^{1/n}$$

$$= \lim_{n\to\infty} \frac{\alpha[1 - (\beta/\alpha)^n]^{1/n}}{(\alpha - \beta)^{1/n}}.$$

Now, $\alpha = (1 + \sqrt{5})/2 = 1.6$ approximately, and $\beta = (1 - \sqrt{5})/2 = -.6$ approximately, and we have $|\beta/\alpha| < 1$, implying $\lim_{n\to\infty} (\beta/\alpha)^n = 0$. Also $\alpha - \beta = \sqrt{5} > 1$, implying $\lim_{n\to\infty} (\sqrt{5})^{1/n} = 1$.

Thus $L = \alpha$.

7. We know that $f_{2n+1} = f_{n+1}^2 + f_n^2$; also $L_n = f_{n+1} + f_{n-1}$ by definition. Therefore we have

$$f_{n+1}L_{n+2} - f_{n+2}L_n = f_{n+1}(f_{n+3} + f_{n+1}) - f_{n+2}(f_{n+1} + f_{n-1})$$

$$= f_{n+1}^2 + f_{n+1}f_{n+3} - f_{n+2}f_{n+1} - f_{n+2}f_{n-1}$$

$$= f_{n+1}^2 + f_{n+1}(f_{n+3} - f_{n+2}) - (f_{n+1} + f_n)(f_{n+1} - f_n)$$

$$= f_{n+1}^2 + f_{n+1}^2 - (f_{n+1}^2 - f_n^2) = f_{n+1}^2 + f_n^2 = f_{2n+1}.$$

Consequently we have

$$S = \sum_{n=1}^{\infty} \frac{f_{2n+1}}{L_n L_{n+1} L_{n+2}} = \sum_{n=1}^{\infty} \frac{f_{n+1}L_{n+2} - f_{n+2}L_n}{L_n L_{n+1} L_{n+2}}$$

$$= \sum_{n=1}^{\infty} \left[\frac{f_{n+1}}{L_n L_{n+1}} - \frac{f_{n+2}}{L_{n+1} L_{n+2}}\right]$$

where each term has the *same form*

$$= \left(\frac{1}{1\cdot 3} - \frac{2}{3\cdot 4}\right) + \left(\frac{2}{3\cdot 4} - \frac{3}{4\cdot 7}\right) + \left(\frac{3}{4\cdot 7} - \frac{5}{7\cdot 11}\right) + \cdots,$$

having partial sums $S_n = 1/3 - (f_{n+2}/L_{n+1}L_{n+2})$.

Since $L_{n+2} > f_{n+2}$, $(f_{n+2}/L_{n+1}L_{n+2}) < 1/L_{n+1}$ which $\to 0$ as $n \to \infty$.

Hence $S = 1/3$.

8. Since $n$ is a given constant, and $\alpha = (1 + \sqrt{5})/2$ is also a constant, we have

$$S = \sum_{k=0}^{n} \binom{n}{k} \alpha^{3k-2n} = \frac{1}{\alpha^{2n}} \sum_{k=0}^{n} \binom{n}{k} (\alpha^3)^k = \frac{1}{\alpha^{2n}} (1 + \alpha^3)^n.$$

Now $\alpha^2 = \alpha + 1$, making

$$\begin{aligned} 1 + \alpha^3 &= 1 + \alpha(\alpha + 1) = 1 + \alpha^2 + \alpha \\ &= (1 + \alpha) + \alpha^2 = 2\alpha^2. \end{aligned}$$

Therefore, $S = (1/\alpha^{2n})(2\alpha^2)^n = (1/\alpha^{2n})\, 2^n \alpha^{2n} = 2^n$.

9. Adding and subtracting the terms corresponding to $k = 0$ and $k = r$, we get

$$\begin{aligned} \sum_{k=1}^{r-1} (-1)^k \binom{r}{k} f_k &= \sum_{k=0}^{r} (-1)^k \binom{r}{k} f_k - (-1)^0 \binom{r}{0} f_0 - (-1)^r \binom{r}{r} f_r \\ &= \left[ \sum_{k=0}^{r} (-1)^k \binom{r}{k} \frac{\alpha^k - \beta^k}{\alpha - \beta} \right] + f_r \quad \text{(since } r \text{ is odd)} \\ &= \frac{1}{\alpha - \beta} [(1 - \alpha)^r - (1 - \beta)^r] + f_r \\ &\qquad\qquad \text{(as argued in part (a))} \\ &= \frac{1}{\alpha - \beta} [(\beta)^r - (\alpha)^r] + f_r \quad \text{(since } \alpha + \beta = 1) \\ &= -\frac{\alpha^r - \beta^r}{\alpha - \beta} + f_r \\ &= -f_r + f_r \\ &= 0. \end{aligned}$$

10. Since

$$f_n = \frac{\alpha^n}{\alpha - \beta} - \frac{\beta^n}{\alpha - \beta}, \quad \text{and} \quad \left| \frac{\beta^n}{\alpha - \beta} \right| < \frac{1}{2},$$

then $f_n$ is the integer nearest $\alpha^n/(\alpha - \beta)$.

Now $\alpha = 1.6180339887\ldots$, and $\log_{10}\alpha = 0.20899\ldots$, $\log_{10} 5 = 0.69897$. Therefore

$$\log_{10}\frac{\alpha^{100}}{\sqrt{5}} = 100(.20899\ldots) - \frac{1}{2}(.69897\ldots)$$

$$= 20.899\ldots - .349\ldots = 20.55\ldots.$$

Hence $f_{100}$ is the integer nearest $10^{20.55\ldots}$, making it a 21-digit integer. The .55 in the mantissa places the first digit at 3.

11. Since $(n^2 - mn - m^2)^2 = 1$, we have $n^2 - mn - m^2 = \pm 1$, $n^2 - mn - (m^2 \pm 1) = 0$, and $n = (m \pm \sqrt{5m^2 \pm 4})/2$, which implies that $5m^2 \pm 4$ must be a perfect square $(y^2)$. By our result in the text, then, for some positive integer $k$, it must be that

$$m = f_k \quad \text{and} \quad y = L_k = f_{k-1} + f_{k+1}.$$

Since $L_k > f_k$ for all $k > 1$, it follows that the positive integer $n$ is given by

$$n = \frac{m + y}{2} = \frac{1}{2}(f_k + f_{k-1} + f_{k+1}) = f_{k+1},$$

making $m$ and $n$ two consecutive Fibonacci numbers. For $m$, $n$ in the range [1,1981], then, the greatest $m^2 + n^2$ is $987^2 + 1591^2$.

# GLOSSARY

**Cauchy Inequality.** For all real numbers $a_i$, $b_i$, $i = 1, 2, \ldots, n$,

$$(a_1^2 + a_2^2 + \cdots + a_n^2)(b_1^2 + b_2^2 + \cdots + b_n^2) \geq (a_1b_1 + a_2b_2 + \cdots + a_nb_n)^2,$$

with equality if and only if $a_i$ and $b_i$ are proportional for all $i$. Reference: G. Pólya and G. Szegö, *Problems and Theorems in Analysis I*, Springer-Verlag, 1974.

**Centroid.** The point of concurrency of the medians of a triangle (a median joins a vertex to the midpoint of the opposite side). It is the center of gravity of a system of equal masses suspended at the vertices.

**Circumcircle.** The circle which passes through all the vertices of a polygon; its center and radius are called the *circumcenter* and *circumradius.*

**Convex Set.** A set of points is convex if, for every two of its points $A$ and $B$, the entire segment $AB$ belongs to the set.

**Convex Hull.** The convex hull $H$ of a set of points $S$ is the "smallest" convex set which contains $S$; $H$ is to be contained in all convex sets that contain $S$, and is, therefore, defined formally to be the intersection of all convex sets that contain $S$. In the case of finite sets, one can think of $H$ as being given by an elastic band that is allowed to contract around $S$.

**Dilatation.** A geometric transformation, sometimes denoted by $O(r)$, which is determined by a center $O$ and a ratio r; $O(r)$ carries a point $P$ to a position in line with $OP$ such that its new distance $OP'$ from $O$ is $r$

times its old distance ($OP'/OP = r$). If $r$ is negative, $OP'$ is layed off along $OP$ on the other side of $O$ (that is, opposite $P$).

**Euler Line.** The centroid, circumcenter, and orthocenter of a triangle always lie on a straight line called the Euler line of the triangle.

**Incircle.** The circle which is tangent to all the sides of a polygon. Its center and radius are called the *incenter* and *inradius.*

**L-Tromino.** The L-shape that results when one square is removed from a set of 4 equal squares that have been put together to form a square.

**Orthocenter.** The point of concurrency of the altitudes of a triangle.

**Pigeonhole Principle.** (A fundamental tool of combinatorics) The principle is the following and refinements thereon: if more than $n$ objects are distributed into a set of $n$ compartments, some compartment must receive more than one of the objects.

**Regular Polygon.** A polygon having equal sides and equal angles.

**Translate.** To move without turning.

# INDEX

Alder, Henry, 68
Alexanderson, G. L., 32
Allaire, Frank, 28
André's Problem, 69
Andrews, George, 39, 45, 68

Baker, Alan, 207
Balasubramanian, 208
Beal, David, 37
Beatty's Theorem, 181
Benkoski, Stan, 215
Bernoulli, Daniel, 109
Bertrand's Postulate, 179, 180, 196
Binet's Formulas, 108, 111
Borwein, D., 220
Brousseau, Alfred, 105
Brown's Criterion, 124
Bruen, A., 220

Carnot's Theorem, 25
Catalan numbers, 146
Cauchy inequality, 3, 4
ceiling function, 225
centroid, 100
Cesaro's observations, 109, 110, 111
Chebyshev polynomials, 208
circular inversion, 29
complementary sequences, 12
complete graph, 60
complete sequence, 123
complex plane, 202, 206
congruent sets, 202, 206
correspondence, 1-1, 45, 67
*Crux Mathematicorum*, 1, 76, 174, 175

diameter, 186
Dijkstra, E. W., 8, 12, 19
Dijkstra-Kluyver, Mrs. B. C., 19
dilatation, 91
Dodge, Clayton, 89

Edwards, Harold, 194
Elder's generalization, 8
Equilic quadrilateral, 32
Erdös, Paul, 37, 146, 196
Euclidean algorithm, 131, 170
Euler, L., 64, 109, 143, 216
Euler line, 100
Euler's Theorem, 65, 169
exponential generating function, 70

Fermat's Theorem, 169
Ferrers graph, 41
Fine, N. J., 45

Gallai, Tibor, 37
Galois Field, 153
Garfunkel, Jack, 32
generating function, 43, 64, 77, 96, 143
Goulden, Ian, 68, 99
Graham, Ron, 89, 129

Hermite, Charles, 207
Hoggatt, Verner Jr., 175
honeybee, 102

Jackson, David, 68
Johnson, Roger, 24
Jordan, J. H., 39

Jung's Theorem, 186
Just, Erwin, 89

Kaplansky, Irving, 146
Kay, David, 27
key (cryptographic), 151
knapsack problem, 163
Koether, Robb, 27
Kürschák's tile, 30

Lagrange, Joseph Louis, 177
Liu, C. L., 61
Lucas, Edward, 110
Lum Wan, Judith, 174

MacMahon, Major Percy, 143
*Mathematical Spectrum*, 37, 176
matrix Q, 106
Michael, Glen, 131
Millin's series, 134
Morsel #9, 21
Morsel #23, 18
Moser, William, 18

Newman, J. R., 193
Niven, Ivan, 177

Old Japanese Theorem, 24
orthocenter, 100

packing problem, 60
partial fractions, 44
partial sums, 146
partitions, 6, 10, 39, 64, 140, 176
pathological set, 205
perfect partitions, 141
permutations, 69
pigeonhole principle, 3, 4, 78, 82, 89, 133, 140
*Pi Mu Epsilon Journal*, 27, 32, 89
Post, K. A., 8, 21
prime numbers, 133, 196
public-key systems, 162
Putnam problem, 1

Ramachandra, 208
reflection, 203
reflection principle, 146
regular polygons, 27
representations, 123
Richmond, Bruce, 208
Richter, Bruce, 180
rotation, 34, 240
RSA system, 166
Russian Olympiad, 2, 80
Ryavec, C., 216

Saari, Donald, 51
Schoenberg, I. J., 32
security, 171
series multisection, 210
Seydel, Kenneth, 32
Shank, Herb, 180
Shapiro, David, 10
Sierpiński, Waclaw, 196
signatures, 167
Singmaster, David, 77
Stanley's Theorem, 6
sum-free sets, 89
Swanson, Norman, 179
sweep line, 18

Taylor's Condition, 58
translation, 37, 203
trees, 60
triangulation, 24
trominos, 85
Turán, Paul, 37

U. S. Olympiad, 28

Vanstone, Scott, 173

Walch, Ray, 39
weakly complete sequence, 129
weighted paths, 57
Wisner, R. J., 39

Yzeren, I. van, 21

Zaks, S., 61
Zeckendorf's Theorem, 129
Zuckerman, Herbert, 177